Statistics for Archaeologists

GW00776243

INTERDISCIPLINARY CONTRIBUTIONS TO ARCHAEOLOGY

Series Editor: Jelmer Eerkens, *University of California Berkeley, Berkeley, CA, USA*
Founding Editor: Roy S. Dickens, Jr. *Late of University of North Carolina, Chapel Hill, NC, USA*

For a complete list of titles in this series, please visit the series online at: http://www.springer.com/series/6090

THE ARCHAEOLOGIST'S LABORATORY
The Analysis of Archaeological Data
E.B. Banning

AURIGNACIAN LITHIC ECONOMY
Ecological Perspectives from Southwestern France
Brooke S. Blades

CASE STUDIES IN ENVIRONMENTAL ARCHAEOLOGY, 2ND EDITION
Elizabeth J. Reitz, Margaret Scarry, and Sylvia J. Scudder

EMPIRE AND DOMESTIC ECONOMY
Terence N. D'Altroy and Christine A. Hastorf

EUROPEAN PREHISTORY: A SURVEY
Edited by Saurunas Miliasuskas

THE EVOLUTION OF COMPLEX HUNTER-GATHERERS
Archaeological Evidence from the North Pacific
Ben Fitzhugh

FAUNAL EXTINCTION IN AN ISLAND SOCIETY
Pygmy Hippotamus Hunters of Cyprus
Alan H. Simmons

A HUNTER-GATHERER LANDSCAPE
Southwest Germany in the Late Paleolithic and Neolithic
Michael A. Jochim

MISSISSIPPIAN COMMUNITY ORGANIZATION
The Powers Phase in Southeastern Missouri
Michael J. O'Brien

NEW PERSPECTIVES ON HUMAN SACRIFICE AND RITUAL BODY
TREATMENTS IN ANCIENT MAYA SOCIETY
Edited by Vera Tiesler and Andrea Cucina

REMOTE SENSING IN ARCHAEOLOGY
Edited by James Wiseman and Farouk El-Baz

THE SCIOTO HOPEWELL AND THEIR NEIGHBORS
Bioarhcoaelogical Documentation and Cultural Understanding
By D. Troy Case and Christopher Carr

THE TAKING AND DISPLAYING OF HUMAN BODY PARTS AS TROPHIES BY AMERINDIANS
Edited by Richard J. Chacon and David H. Dye

Statistics for Archaeologists

A Common Sense Approach

Second Edition

Robert D. Drennan

 Springer

Dr. Robert D. Drennan
University of Pittsburgh
Dept. Anthropology
Pittsburgh PA 15260
USA
drennan@pitt.edu

ISSN 1568-2722
ISBN 978-1-4419-0412-6 (hardcover)   e-ISBN 978-1-4419-0413-3
ISBN 978-1-4419-6071-9 (softcover)
DOI 10.1007/978-1-4419-0413-3
Springer New York Dordrecht Heidelberg London

Library of Congress Control Number: 2010922892

Printed on acid-free paper

Springer is part of Springer Science+Business Media (www.springer.com)

Preface to the Second Edition

This book is intended as an introduction to basic statistical principles and techniques for the archaeologist. It grows primarily from my experience in teaching courses in quantitative analysis for undergraduate and graduate students in archaeology over a number of years. The book is set specifically in the context of archaeology, not because the issues dealt with are uniquely archaeological in nature, but because many people find it much easier to understand quantitative analysis in a familiar context – one in which they can readily understand the nature of the data and the utility of the techniques. The principles and techniques, however, are all of much broader applicability. Physical anthropologists, cultural anthropologists, sociologists, psychologists, political scientists, and specialists in other fields make use of these same principles and techniques. The particular mix of topics, the relative emphasis given them, and the exact approach taken here, however, do reflect my own view of what is most useful in the analysis of specifically archaeological data.

It is impossible to fail to notice that many aspects of archaeological information are numerical, and that archaeological analysis has an unavoidably quantitative component. Standard statistical approaches are commonly applied in straightforward as well as unusual and ingenious ways to archaeological problems, and new approaches have been invented to cope with the special quirks of archaeological analysis. The literature on quantitative analysis in archaeology has grown to prodigious size. Some of this literature is extremely good, while some of it reveals only that publishing on statistics in archaeology is an activity open even to those whose comprehension of the most fundamental statistical principles is primitive at best. The article attempting to point out which published work fits into which of these categories has itself become a recognizable genre. This book does not attempt to evaluate or criticize in such a mode, but it is motivated in part by the perception that, as a group, those of us responsible for training archaeologists in quantitative analysis can claim only mixed success to date. Consequently, this book is in part a discussion of how quantitative data analysis is done in archaeology but in larger part a discussion of how quantitative data analysis *could be* done in archaeology. Its focus is resolutely on some fundamental principles and how they can be applied most usefully in archaeology. It is tempting to discuss the numerous variations in these applications that might be made in analyzing archaeological data and to provide examples of ways in which these principles have actually been put to work by archaeologists. I have, however, attempted to resist these temptations in an effort to keep the focus firmly on basic principles and to provide brief and clear explanations of them. It is to maintain simplicity and clarity that both the examples used in the text and the practice problems at the ends of the chapters are made up rather than selected from real archaeological data. I assume that the readers of this book know enough about archaeology not to need descriptions and pictures of post holes, house floors, scrapers, or sherds – that we all know what it means to say that we have conducted a regional survey and measured the areas of 53 sites.

Most of the techniques in this book are fairly standard, either in the "classical" statistics developed between 1920 and 1950 or in the more recent "exploratory data analysis" school. The approach or, perhaps more important, the general attitude of this book derives ultimately from the work of John W. Tukey and his colleagues and students, progenitors of exploratory data analysis, or EDA for short. As is usual in general books on statistics, I have not included bibliographic citations in the text, but Suggested Reading appears at the end. This book leans toward the terminology of EDA, although the equivalent more traditional terms are usually mentioned. Where it makes the explanations easier to understand in the context of archaeology, the terminology used here is simply nonstandard.

Archaeologists (and others) sometimes are as wary of statistics as school children are of the classroom holding the most imposing disciplinarian among the teachers. Statistics seems a place filled with rules the rationale of which is opaque, but the slightest infraction of which may bring a painful slap across the knuckles with a ruler. This attitude has certainly been reinforced by critiques that take published work in archaeology to task for breaking sacred statistical rules. It may come as a surprise to many to learn that a number of conflicting versions exist of many statistical rules. Statisticians, like the practitioners of any other discipline, often disagree about what are productive approaches and legitimate applications. Use of statistical tools often involves making subjective judgments. In an effort to provide a sound basis for such judgments, introductory texts often attempt to reduce them to clear-cut rules, thereby creating considerable confusion about what are really fundamental principles and what are merely guidelines for difficult subjective decisions.

In short, the rules of statistics were not on the stone tablets Moses brought down from the mountain. This book openly advocates the overthrow of rules found in some texts (by reason and common sense rather than force and violence). Since it is intended as an introduction to statistical principles, long arguments against alternative approaches are not appropriate. One issue, however, is of such central importance that it must be mentioned. The approach taken to significance testing here does not involve rigid insistence on either rejecting or failing to reject a "null hypothesis." In archaeology it is much more informative in most instances simply to indicate how likely it is that the null hypothesis is correct. The rigorous formulation of the null hypothesis, then, does not get the all-consuming attention here that is sometimes devoted to it elsewhere. In this approach to significance testing and to several issues related to sampling, I have followed the lead of George Cowgill (see

Suggested Reading at the end of the book), although I have not carried into practice all of the thoroughly sensible suggestions he has made. (One obstacle to following some of his suggestions continues to be, as he noted, that few of the available statistics computer programs provide the necessary information in their output.) To those who were taught that significance testing was built upon the rock of rejecting or failing to reject the null hypothesis, I recommend thoughtful attention to the points Cowgill makes.

The approach taken to significance testing makes clear thinking about populations, samples, and sampling procedures especially important. Indeed, in many contexts, it makes simply using samples to make statements about the populations they came from a more appealing approach than significance testing. It is for this reason that samples and sampling are given much lengthier treatment here than is common in introductory books on statistics. Part I of this book is about exploring batches of numbers in ways that are interesting and useful in and of themselves, but that are especially chosen for their relevance when batches are considered samples from larger populations. Part II develops this notion of batches as samples and makes a frontal assault on some of the central principles that relate samples to populations. Part III presents a fairly standard suite of basic tests of the strength and significance of relationships between two variables, together with alternative approaches derived directly from sampling estimation. Part IV returns to take up a series of separate issues related to sampling – issues of special importance in archaeology. These chapters relate most directly to those in Part II, but they have been placed later on to avoid interrupting the steady progression of ideas that links Parts II and III. Finally, Part V attempts a rapid introduction to exploring multivariate datasets for patterning. It brings us back to the exploratory data analysis attitude most strongly reflected in Part I.

In archaeology, as in most fields, quantitative concepts come easily and naturally to some, and only at considerable cost to others. The absence of a natural inclination toward numerical reasoning is often reinforced by the social acceptability of professing ignorance of mathematics – a social acceptability nurtured by the notion that mathematics is an arcane and specialized subject of no use to very many people. An otherwise well-educated person can profess a complete inability to comprehend anything about numbers beyond addition and subtraction without incurring the disdain to be expected if he or she admitted to verbal skills so limited as to make everything in the daily newspaper but the comics unintelligible.

Varying degrees of natural talent should be no more surprising for mathematics than for writing, playing football, or other activities. The view that mathematics is only a necessary evil of elementary school, however, aggravates the problem by encouraging those who have found quantitative reasoning difficult to minimize its importance and to avoid developing quantitative skills that would be useful to them. Consequently, a good many students seem to embark on graduate study of archaeology equipped only with high school algebra – victims, perhaps, of the same kind of bad advice I myself received as a first-semester freshman in college, when my academic advisor scornfully dismissed the math course I intended to enroll in as irrelevant to my interests.

This book is written in the hope of providing useful tools for quantitative analysis in archaeology to those naturally adept at quantitative reasoning as well as to those who find mathematics not only difficult but even intimidating. It is no challenge to present statistics to those already comfortable with and adept at mathematical thinking; it requires only a nudge in the right direction. The perennial challenge of books such as this, however, is to present quantitative analysis effectively to those to whom it does not come naturally. It is with particular concern for this latter group that the approach taken here was chosen. Part of that approach is to plunge right ahead to the tools this book is about without a series of preliminary chapters laying basic groundwork, the importance of which only becomes apparent later on. These "basics" are, instead, discussed as briefly as possible at the points where they become relevant.

Fortunately, it is possible to approach basic statistical tools with common sense and in common language so as to convey not only the mechanics of using the tools of statistics but also a genuine understanding of the way the tools work. Productive use of statistical tools in archaeology springs not so much from abstract mathematical knowledge as from solid intuitive understanding of principles, applied with common sense and unwavering attention to the final product desired – that is, the ultimate research objective. It is worth pausing to emphasize that this book, fundamentally, is about tools – tools for identifying patterns in numbers and tools for assessing how precisely and how reliably the patterns we identify in our data represent real patterns in the broader world our conclusions really are about. As with carpenters' tools, for example, skillful use of statistical tools does not require complete knowledge of how the tools are made. Consequently, I have not attempted to show how statistical equations are derived from certain assumptions through mathematical logic (the approach followed by some books on statistics). As powerful and elegant as the language of abstract mathematics may be, it remains utterly impenetrable to many archaeologists. I have always found it helpful to avoid an abstract mathematical approach. This seems especially important to those already frightened at the thought of mathematics.

Although learning to use a table saw does not require developing the ability to make one, skillful use of a table saw does require some understanding of the principles according to which it does its work. Failure to understand these basic principles will lead to erroneous and uneven cutting and even the occasional severed finger or worse. In just the same way, skillful use of statistical tools requires true understanding of underlying principles. Without such understanding, even very keen statistical tools produce only crude results, and they can cause injury (although generally not the kind that requires medical attention).

For this reason, I have also tried to avoid the cookbook approach common to books on applied statistics. Easy recipes for statistical analysis appeal strongly, especially to those afraid of mathematics. No real mental labor seems to be required; no difficult concepts need be mastered; just carefully follow the instructions. This approach may actually work in disciplines where certain kinds of data are regularly produced in certain formats. Only the most routine data analysis tasks can be successfully handled in this manner, however, and archaeological data are never routine. The nature of the archaeological record and the manner in which we must extract data from it inevitably produce idiosyncrasies that practitioners in other disciplines are taught to avoid through appropriate research design. Coping with such messy data requires that the archaeologist have a better grasp of underlying principles than a cookbook approach can provide.

This book, then, seeks a middle ground. It attempts more than simply providing instructions for the use of statistical tools; yet it makes no pretense of providing a complete mathematical justification for them. Its aim is to help the reader understand the principles underlying statistical tools well enough to use them skillfully in the context of archaeological data analysis. The reader I had in mind while writing is primarily the graduate or undergraduate student of archaeology taking a first course in archaeological data analysis. Like most textbooks, this is the book the author always wanted but never found for his own course. I hope it may also be useful to archaeologists who wish to develop or consolidate skills in statistical tool use whether they are enrolled in courses or not.

The statistical tools discussed in this book by no means make up the complete set ever needed by the archaeologist. They are basic general-purpose tools, but many other specialized tools exist. Some of the tools presented here are quite simple and easy to apply, requiring nothing more than pencil and paper or perhaps an ordinary calculator. Others are more complicated or involve very cumbersome calculations. I take it for granted that any serious archaeological data analysis effort will be undertaken with the aid of a computer. Learning to use statistical software packages is best incorporated directly into the process of learning about the statistical tools. I have thus omitted the often time-consuming and complex explanations of how to compute certain complicated statistics by hand. While calculating some things out by hand can facilitate understanding, one soon reaches the point where preoccupation with the mechanics of calculations interferes with attention that should be devoted directly to underlying principles.

Many of the results and examples in this book were produced with SYSTAT$^{\circledR}$; other packages that could be used are too numerous even to list. Since the possibilities are so varied (and change so continually), it is useless to attempt to incorporate instructions for using statistical software into this book. I assume, however, that the book will be used in conjunction with some package of statistical programs and the corresponding manuals, and some general comments about using such "statpacks" are included.

Almost any software package will provide options and choices not discussed in this book. Some software manuals provide good explanations of what these options are and bibliographic citations for those interested in learning more about them; other manuals do not. (This is one feature worth weighing in choosing statistical software.) Serendipitous encounters with options in statistical software can provide a useful means of expanding one's expertise in quantitative analysis. On the other hand, they can distract the analyst's attention from the task at hand to the many other tasks that could be performed but that there is really no need to perform. The professional carpenter does not first choose a pretty tool and then go looking for something to use it on. Just so, the skilled data analyst first determines what analysis to perform and then turns to pencil, paper, calculator, or computer (as may be appropriate) to put into use the appropriate tool to accomplish the task at hand. The mechanics of complicated calculations and complicated computer software can both divert attention away from central matters of principle concerning the work to be done. In statistics, as in the several sports from which the cliché is derived, it is impossible to remind yourself too often to keep your eye on the ball.

ACKNOWLEDGEMENTS

The person most responsible for "infecting" me (his word, not mine) with the attitude toward statistics represented here is Lee Sailer. Mark Aldenderfer and Doug Price provided very helpful reactions to the manuscript of the first edition. I have stubbornly refused to accept some of the advice generously offered by all three, however, so they cannot be blamed for any deficiencies. Jeanne Ferrary Drennan has put up with a lot of cursing as I tried to teach courses in archaeological data analysis with texts I didn't like, and she dedicated most of one December vacation to helping put the first draft of this book in shape to use in class in January. She pitched in once again with vital help in getting the manuscript for the second edition ready to send to the publisher, as did Adam Menzies and Scott Palumbo. My most special thanks are reserved for the graduate and undergraduate students (and the teaching assistants) who have struggled gamely along as I tried to give enough coherence to this approach to data analysis in archaeology to use it in the courses they took – sometimes using texts it contradicted, sometimes using no text at all, and finally using successive versions of this book. They have contributed more than they know to whatever clarity the exposition here may have.

Pittsburgh, PA   Robert D. Drennan

Contents

Part I
Numerical Exploration

Chapter 1
Batches of Numbers

A *batch* is a set of numbers that are related to each other because they are different instances of the same thing. The simplest example of a batch of numbers is a set of measurements of different examples of the same kind of thing. For example, the lengths of a group of scrapers, the diameters of a group of post holes, and the areas of a group of sites are three batches of numbers. In these instances, length, diameter, and area are *variables* and each scraper, post hole, and site is a *case*.

The length of one scraper, the diameter of one post hole, and the area of one site do not, together, make a batch of numbers because they are completely unrelated. The length, width, thickness, and weight of one scraper do not, together, make a batch because they are not different instances of the same thing; that is, they are different variables measured for a single case. The length, width, thickness, and weight of each of 20 scrapers make, not one batch of numbers, but four. These four batches can be related to each other because they are four variables measured for the same 20 cases. The diameters of a set of 18 post holes from one site and the diameters of a set of 23 post holes from another site can be considered a single batch of numbers (the variable *diameter* measured for 41 cases, ignoring entirely which site each post hole appeared in). They can also be considered two related batches of numbers (the variable *diameter* measured for 18 cases at one site and 23 cases at another site). Finally they can be considered two related batches of numbers in a different way (the variable *diameter* measured for 41 cases and the variable *site* classified for the same 41 cases). This last, however, carries us to a different kind of batch or variable, and it is easier to stick to batches of measurements for the moment.

R.D. Drennan, *Statistics for Archaeologists*, Interdisciplinary Contributions
to Archaeology, DOI 10.1007/978-1-4419-0413-3_1,
© Springer Science+Business Media, LLC 2004, 2009

STEM-AND-LEAF PLOTS

A list of measurements does not lend itself very well to making interesting observations, so the first step in exploration of a batch of numbers is to organize them. If
the batch is a set of measurements, the *stem-and-leaf plot* is the fundamental organ-
izational tool. Consider the batch of numbers in Table 1.1. Ordering them along a
scale can often help us to see patterns. Figure 1.1 shows how to produce a stem-
and-leaf plot that does exactly this for the numbers in Table 1.1. First, the numbers
are divided into a stem section and a leaf section. In the first case, for instance, 9.7
becomes a stem of 9 and a leaf of 7. The leaf for each number is placed on the stem
plot beside the stem for that number. The lines in Fig. 1.1 connect some of the numbers to the corresponding leaves in their final positions on the stem-and-leaf plot.
(Not all the connections are drawn in to avoid a hopeless confusion of lines.)

Several characteristics of this batch of numbers are immediately apparent in the
stem-and-leaf plot. First, the numbers tend to bunch together at about 9 to 12 cm.
Most fall in this range. Two more (14.2 and 7.6 cm) fall a little outside this range,
and one (44.6 cm) falls far away from the rest. It is a fairly common occurrence for
batches of numbers to bunch together like this. It is also relatively common for one
or a few numbers in a batch to fall far away from the bunch where the majority of
the numbers lie. Such numbers that fall far from the bunch are often called *outliers*,
and we will discuss them in more detail later. For now it is sufficient to note that we
often examine such outliers with a skeptical eye. A post hole 44.6 cm in diameter
is certainly a very unusual post hole in this batch, and we might be suspicious that
someone has simply written the measurement down wrong. A quick check of field
drawings or photographs should be sufficient to determine whether such an error
has been made and, if so, to correct it. If, indeed, this measurement seems correct,
then one of the conspicuous features of this batch is that one post hole simply does
not seem to fit with the rest of the group.

Stem-and-leaf plots can be made at different scales (that is, using different intervals on the stem), and the selection of an appropriate scale is essential to producing a
helpful stem-and-leaf plot. Table 1.2 shows another batch of numbers in a stem-and-
leaf plot at the same scale as in the previous example. The numbers here, however,
are spread out over such a large distance that the characteristics of the batch are not

**Table 1.1. Diameters of 13
Post holes at the Black
Site (cm)**

9.7	11.7
9.2	11.1
12.9	7.6
11.4	11.8
9.1	14.2
44.6	10.8
10.5	

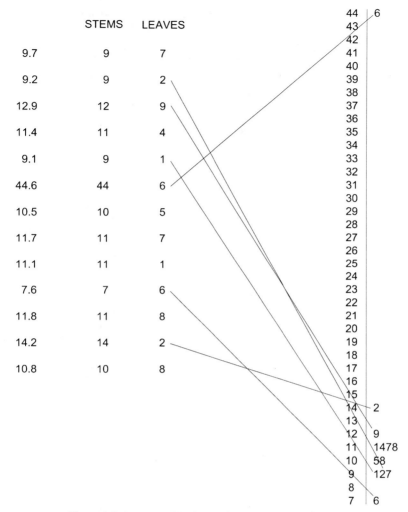

Figure 1.1. A stem-and-leaf plot of the numbers in Table 1.1.

clearly displayed. In Table 1.3 the same numbers yield a denser stem-and-leaf plot
when the stem is structured differently. In the first place, the numbers are broken
differently into stem and leaf sections – not at the decimal point but between the
units and tens. Since there are two digits for each leaf, commas are used to indicate the separation between leaves. To avoid greatly increasing the density, two
positions are allowed on the stem for each stem section, the lower position corresponding to the lower half of the numbers that might fit that stem section and
the upper corresponding to the upper half (as indicated by the notations to the right
of the stem-and-leaf plot). The characteristics of the batch are much clearer in this
plot. The numbers bunch together from about 130 to 160. And one unusually light

**Table 1.2. Too Sparse Stem-and-Leaf Plot of Weights
of 17 Scrapers from the Black Site**

Weight (g)	Stems	Leaves
148.7	148	7
154.5	154	5
169.5	169	5
145.1	145	1
157.9	157	9
137.8	137	8
151.9	151	9
146.2	146	2
164.7	164	7
149.3	149	3
141.3	141	3
161.2	161	2
146.9	146	9
152.0	152	0
143.0	143	0
132.6	132	6
115.3	115	3

Stem	Leaf
169	5
168	
167	
166	
165	
164	7
163	
162	
161	2
160	
159	
158	
157	9
156	
155	
154	5
153	
152	0
151	9
150	
149	3
148	7
147	
146	29
145	1
144	
143	0
142	
141	3
140	
139	
138	
137	8
136	
135	
134	
133	
132	6
131	
130	
129	
128	
127	
126	
125	
124	
123	
122	
121	
120	
119	
118	
117	
116	
115	3

**Table 1.3. Stem-and-Leaf Plot at an Appropriate Scale of Weights of 17
Scrapers from the Black Site**

Weight (g)	Stems	Leaves			
148.7	14	87			
154.5	15	45			
169.5	16	95	17		(175.0–179.9)
145.1	14	51	17		(170.0–174.9)
157.9	15	79	16	95	(165.0–169.9)
137.8	13	78	16	12,47	(160.0–164.9)
151.9	15	19	15	79	(155.0–159.9)
146.2	14	62	15	19,20,45	(150.0–154.9)
164.7	16	47	14	51,62,69,87,93	(145.0–149.9)
149.3	14	93	14	13,30	(140.0–144.9)
141.3	14	13	13	78	(135.0–139.9)
161.2	16	12	13	26	(130.0–134.9)
146.9	14	69	12		(125.0–129.9)
152.0	15	20	12		(120.0–124.9)
143.0	14	30	11	53	(115.0–119.9)
132.6	13	26			
115.3	11	53			

scraper seems to be an outlier. This pattern can certainly be detected (especially in hindsight) in Table 1.2, but it is much clearer in Table 1.3.

Table 1.4 shows a still denser stem-and-leaf plot of the same numbers. Stem and leaf sections are separated as in Table 1.4, but only one position is allowed on the stem for each stem section. At this scale, the bunching of numbers is still evident, but what seemed an outlier in Table 1.4 has come so close to the bunch that it no longer seems very different. The characteristics of the batch are less clearly displayed in this stem-and-leaf plot because it crowds the numbers too closely together.

Table 1.5 is yet another stem-and-leaf plot of the same numbers. This one is much too dense. There is simply not enough room on the stem for the leaves to spread out far enough to show the patterning. The outlier from Table 1.3 is no longer apparent (although it is still there – it is just obscured by the inappropriate scale). It is difficult even to evaluate the extent of the bunching of numbers. You can create the next step in the direction of denser stem-and-leaf plots for these numbers yourself. It has a stem consisting only of 1, with all the leaves in one line next to it.

An appropriate scale for a stem-and-leaf plot avoids the two extremes seen in Tables 1.2 and 1.5. The leaves should make one or more branches or bunches of leaves that protrude from the stem. This cannot happen if they are spread out along a stem that is simply too long as in Table 1.2. At the same time, the leaves should be allowed to spread out enough so that outliers can be noticed and two or more bunches, if they occur, can be distinguished from one another. This latter cannot happen if the leaves are crowded together as in Table 1.5. Tables 1.3 and 1.4 show stem-and-leaf plots at scales that are clearer, although Table 1.3 definitely shows the patterns more clearly than Table 1.4 does.

**Table 1.4. Too Dense a Stem-and-Leaf Plot of Weights of 17 Scrapers
from the Black Site**

Weight (g)	Stems	Leaves		
148.7	14	87		
154.5	15	45		
169.5	16	95		
145.1	14	51	17	
157.9	15	79	16	12,47,95
137.8	13	78	15	19,20,45,79
151.9	15	19	14	13,30,51,62,69,87,93
146.2	14	62	13	26,78
164.7	16	47	12	
149.3	14	93	11	53
141.3	14	13		
161.2	16	12		
146.9	14	69		
152.0	15	20		
143.0	14	30		
132.6	13	26		
115.3	11	53		

Different statisticians make stem-and-leaf plots in slightly different ways. There are several approaches to spreading out or compressing the scale. The exact format followed is less important than to show as clearly as possible the patterns to be observed in the batch of numbers. Two essential principles are involved. First, the distances between the numbers are represented visually as spatial distances along the vertical number scale in the graph. And second, the number of numbers in each of a series of equal intervals is represented visually as a spatial distance along each horizontal row of numbers. However the stem sections are divided, it is important that each stem section correspond to a range of numbers equal to that of every other stem section. It would be a bad idea to structure a stem with positions corresponding to, say, 3.0–3.3, 3.4–3.6, and 3.7–3.9 because the intervals are unequal. That is, a larger range is included between 3.0 and 3.3 than in the other two intervals. There will tend to be longer rows of leaves for that larger interval, simply because it is a larger interval, and that interferes with the horizontal spacing principle that enables the stem-and-leaf plot to do its work.

The stem-and-leaf plots in this book have lower numbers at the bottom and higher numbers at the top. This makes it easier to talk about numbers and stem-and-leaf plots in the same terms since lower numbers are lower on the plot and higher numbers are higher on the plot. It is more common for stem-and-leaf plots to be drawn with lower numbers at the top and higher numbers at the bottom. This is unfortunate because it adds a small and entirely unnecessary element of confusion, but either way, the stem-and-leaf plot shows the same patterns.

Finally, the stem-and-leaf plots in the tables in this chapter have the leaves on each line in numerical order. This makes no difference in observing the kinds of

Table 1.5. Much Too Dense a Stem-and-Leaf Plot of Weights of 17 Scrapers from the Black Site

Weight (g)	Stems	Leaves		
148.7	1	487		
154.5	1	545		
169.5	1	695		
145.1	1	451		
157.9	1	579		
137.8	1	378		
151.9	1	519		
146.2	1	462	1	519,520,545,579,612,647,695
164.7	1	647	1	153,326,378,413,430,451,462,469,487,493
149.3	1	493		
141.3	1	413		
161.2	1	612		
146.9	1	469		
152.0	1	520		
143.0	1	430		
132.6	1	326		
115.3	1	153		

Table 1.6. Diameters of 15 Post holes at the Smith Site (cm)

20.5	19.4
17.2	16.4
15.3	18.8
15.9	15.7
18.3	18.9
17.9	16.8
18.6	8.4
14.3	

patterns we have been noting here, but it does make it easier to do some of the things we will do with stem-and-leaf plots in Chapters 2 and 3. It makes drawing a stem-and-leaf plot a little more time consuming, but it is well worth the effort, as we shall see.

BACK-TO-BACK STEM-AND-LEAF PLOTS

The stem-and-leaf plot is a fundamental tool not just for exploring a single batch but also for comparing batches. The batch of numbers in Table 1.6 consists of post hole diameters from the Smith Site, which we may want to compare to the batch of post hole diameters from the Black Site (Table 1.1). These batches can be related

Table 1.7. Back-to-Back Stem-and-Leaf
Plot of Post hole Diameters from the Black
and Smith Sites (Tables 1.1 and 1.6)

Black Site		Smith site
6	44	
	43	
	42	
	41	
	40	
	39	
	38	
	37	
	36	
	35	
	34	
	33	
	32	
	31	
	30	
	29	
	28	
	27	
	26	
	25	
	24	
	23	
	22	
	21	
	20	5
	19	4
	18	3689
	17	29
	16	48
	15	379
2	14	3
	13	
9	12	
8741	11	
85	10	
721	9	
	8	4
6	7	

since they are measurements of the same variable (diameter of post holes), although two different sets of post holes are involved. Table 1.7 shows a *back-to-back stem-and-leaf plot* in which the leaves representing both batches of numbers are placed on opposite sides of the same stem.

We see the bunch of post holes at diameters of 9–12 cm that we saw for the Black Site in Fig. 1.1, as well as the outlier, or unusually large post hole 44.6 cm in diameter. For the Smith Site we see a bunch of numbers as well, but this bunch of numbers falls somewhat higher on the stem than the bunch for the Black Site. We quickly observe, then, that the post holes at the Smith Site are in general of larger diameter than those at the Black Site. This general pattern is unmistakable in the stem-and-leaf plot even though the 44.6-cm post hole at the Black Site is by far the largest post hole in either site. There is also an outlier among post holes at the Smith Site – in this instance a low outlier much smaller than the general run of post holes at the site. If this post hole were at the Black site instead of the Smith Site, it would not be nearly so unusual, but at the Smith Site it is clearly a misfit.

HISTOGRAMS

The stem-and-leaf plot is an innovation of exploratory data analysis. Although it has certainly appeared in the archaeological literature, there is a traditional way of drawing plots with similar information that is probably more familiar to more archaeologists. It is the *histogram*, and it corresponds precisely to the stem-and-leaf plot. The histogram is familiar enough that no detailed explanation of it is needed here. Table 1.8 provides a stem-and-leaf plot of the areas of 29 sites in the Kiskiminetas River Valley. Figure 1.2 shows that a histogram of this same batch of numbers is simply a boxed-in stem-and-leaf plot turned on its side with the numbers themselves eliminated as leaves. Most of the same patterns we have noted up to now in stem-and-leaf plots can be observed in histograms as well. In making a histogram, one faces the same choice of scale or interval that we have already discussed for the stem-and-leaf plot, and precisely the same considerations apply. Histograms have the advantage of being somewhat more elegant and esthetically pleasing as well as of being more familiar to archaeologists. Stem-and-leaf plots, on the other hand, have the advantage that the full detail of the actual numbers is all present, and this makes it possible to use them in ways that histograms cannot be used, as we shall see in Chapters 2 and 3. In general terms, however, the stem-and-leaf plot and the histogram serve fundamentally the same purpose.

MULTIPLE BUNCHES OR PEAKS

The batch of numbers in Table 1.8 also demonstrates another characteristic of batches that sometimes becomes obvious in either a stem-and-leaf plot or a histogram. We see the usual bunching of numbers in the stem-and-leaf plot. In this case, however, there are two distinct and separate bunches, one between about 1 and 5 ha and another between about 7 and 16 ha. The same bunches are obvious in the histogram (Fig. 1.2), where the two separate bunches appear as two hills or

**Table 1.8. Areas of 29 Sites in the
Kiskiminetas River Valley**

Site area (ha)	Stem-and-leaf plot	
12.8	15	3
11.5	14	0
14.0	13	49
1.3	12	388
10.3	11	0257
9.8	10	367
2.3	9	089
15.3	8	27
11.2	7	4
3.4	6	
12.8	5	
13.9	4	5
9.0	3	48
10.6	2	0239
9.9	1	37
13.4		
8.7		
3.8		
11.7		
1.7		
12.3		
11.0		
2.9		
10.7		
7.4		
8.2		
2.0		
2.2		
4.5		

peaks. Such a pattern of multiple bunches or peaks is a clear indication of distinct kinds of cases – in this instance two distinct kinds of sites. We might likely call them large sites and small sites, and the pattern seen in the stem-and-leaf plot or the histogram indicates that the two are clearly separate. That is, in discussing these as large and small sites, we would not be arbitrarily dividing sites up into large and small but rather responding to an innate characteristic of this batch of numbers. We see quickly that the large sites are more numerous, but there are enough small sites to form a clear and separate peak. This is not a case of outliers but instead, of two sets of sites, each numerous enough to form its own peak in the histogram.

The presence of multiple peaks in a batch is always an indication that two or more fundamentally different kinds of things have been thrown together and measured. To take a ridiculous example, I might measure the diameters of a series of dinner plates and manhole covers. If I presented these as a single list of measurements of

Figure 1.2. A histogram of areas of 29 sites in the Kiskiminetas River Valley.

round objects, you would see immediately in a stem-and-leaf plot that there were two separate peaks. Knowing nothing about the objects except their diameters, you would guess that two fundamentally different kinds of things had been measured. You would be correct to subdivide the batch into two batches with no further justification than the pattern you saw in the stem-and-leaf plot. One of the first things you might do, however, would be to seek further information about the nature of the objects that might clarify their differences. Your reaction, on finding out that both dinner plates and manhole covers were included among the objects measured, might well be "No wonder; now I understand!" This is a perfectly appropriate reaction and would put substance behind a division made on purely formal grounds (that is, on the basis of the pattern observed in a stem-and-leaf plot).

To repeat, batches with multiple peaks cannot be analyzed further. The only correction for this problem is to subdivide the batch into separate batches for separate analysis. In the best of all possible worlds, we can identify other characteristics of the objects in question to aid us in making the division. If not, we must do it simply on the basis of the stem-and-leaf plot or histogram, drawing a dividing line on the number scale at the lowest point of the valley that separates the peaks. This is especially easy for the numbers illustrated in Fig. 1.2. The lowest point of the valley here is around 6 ha. There are no sites at all of this size, so the small sites are clearly those ranging from 1 to 5 ha, and the large sites are those ranging from 7 to 16 ha. If there is not an actual gap at the bottom of the valley, as there is in this instance, just where to draw the dividing line may not be so obvious, but it must be done nevertheless before proceeding to any further analysis.

Statpacks

The stem-and-leaf plot is such a simple way to display the numbers in a batch that it can be produced quickly and easily with pencil and paper. When working with pencil and paper, it is necessary only to be careful to line the numbers up vertically so that the patterns are represented accurately. It is also easy to use a word processor to produce a stem-and-leaf plot. As when working with pencil and paper, it is important to line the numbers up vertically. This happens automatically as long as the font chosen shows all characters (or at least all numbers) as the same width. Fonts in which 1, for example, is narrower than 2 don't work for stem-and-leaf plots because the numbers will get out of alignment. The easiest way to make stem-and-leaf plots, of course, is with a statistics computer package, or *statpack* for short. A statpack will perform the entire operation automatically, including choosing an appropriate scale or interval for the stem. Some statpacks still do not include exploratory data analysis (EDA) tools like stem-and-leaf plots, but many do.

Histograms are more time consuming to draw nicely than stem-and-leaf plots, but many statpacks do a very good job of it. True statistical packages are best for this task, since their programmers had in mind exactly the goals discussed in this chapter when they wrote the programs. Numerous programs that draw bar graphs might at first glance seem another option, but bar graphs, while superficially similar to histograms, are actually a different tool—one that we will explore more fully in Chapter 6.

PRACTICE

In Tables 1.9 and 1.10 are two batches of numbers – measurements of the lengths of scrapers recovered from two sites. The scrapers are made from either flint or chert. These numbers could be considered a single batch of numbers (lengths of scrapers, disregarding what raw material they were made from and what site they

Table 1.9. Scrapers from Pine Ridge Cave

Raw material	Length (mm)	Raw material	Length (mm)
Chert	25.8	Chert	25.9
Chert	6.3	Chert	23.8
Flint	44.6	Chert	22.0
Chert	21.3	Chert	10.6
Flint	25.7	Flint	33.2
Chert	20.6	Chert	16.8
Chert	22.2	Chert	21.8
Chert	10.5	Flint	48.3
Chert	18.9		

Table 1.10. Scrapers from the Willow Flats Site

Raw material	Length (mm)	Raw material	Length (mm)
Chert	15.8	Flint	49.1
Flint	39.4	Flint	41.7
Flint	43.5	Chert	15.2
Flint	39.8	Chert	21.2
Chert	16.3	Flint	30.2
Flint	40.5	Flint	40.0
Flint	91.7	Chert	20.2
Chert	21.7	Flint	31.9
Chert	17.9	Flint	42.3
Flint	29.3	Flint	47.2
Flint	39.1	Flint	50.5
Flint	42.5	Chert	10.6
Flint	49.6	Chert	23.1
Chert	13.7	Flint	44.1
Chert	19.1	Flint	45.8
Flint	40.6		

came from). They also form two related batches in two different ways. We could divide the single batch into two batches according to which site the scrapers were recovered at. (This is the way the numbers are presented in the tables.) Or we could divide the single batch into two batches according to which raw material they were made of (disregarding which site they came from).

1. Make a stem-and-leaf plot of scraper lengths, treating the entire set of scrapers as a single batch. Experiment with different intervals for the stem to consider which interval produces the most useful plot. What patterns do you see in the plot?
2. Make a back-to-back stem-and-leaf plot of scraper lengths, treating the scrapers from the Willow Flats site as one batch and those from Pine Ridge Cave as another batch. (That is, ignore the raw material of which the scrapers were made for the moment.) How do the two batches compare to each other? Do you see any patterns that help you interpret the stem-and-leaf plot of all scrapers as a single batch?
3. Make a back-to-back stem-and-leaf plot of scraper lengths, treating the flint scrapers as one batch and the chert scrapers as another batch. (That is, this time ignore which site the scrapers came from.) How do these two batches compare to each other? Do you see any patterns this time that help you interpret the stem-and-leaf plot of all scrapers as a single batch?

Chapter 2
The Level or Center of a Batch

As we saw in Chapter 1, the numbers in a batch often bunch together. If we compare two related batches of numbers, the principal bunch in one batch may well have higher numbers in general than the principal bunch in the other batch. We say that such batches have different *levels* or *centers*. It is convenient to use a numerical index of the level for such comparisons. The several such indexes in common use are traditionally referred to as *measures of central tendency*.

THE MEAN

The most familiar index of the center of a batch is the *mean*, outside statistics more commonly referred to as the *average*. Calculation of the mean is just as we all learned in elementary school: the sum of all the numbers in the batch is divided by the number of numbers in the batch. Since this is such a familiar calculation, it provides a good opportunity to introduce some mathematical notation that is particularly useful in statistics. The equation expressing the calculation of the mean is

$$\overline{X} = \frac{\Sigma x}{n}$$

where x represents each number in a batch, individually, n is the number of x's, and $\overline{X}$ is the mean or average of x (pronounced "x bar").

R.D. Drennan, *Statistics for Archaeologists*, Interdisciplinary Contributions
to Archaeology, DOI 10.1007/978-1-4419-0413-3_2,
© Springer Science+Business Media, LLC 2004, 2009

Table 2.1. Weights of Flakes Recovered from Two Bell-Shaped Pits

	Flake weights (g)		Back-to-back stem-and-leaf plot		
	Pit 1	Pit 2	Pit 1		Pit 2
	9.2	11.3	6	28	
	12.9	9.8		27	
	11.4	14.1		26	
	9.1	13.5		25	
	28.6	9.7		24	
	10.5	12.0		23	
	11.7	7.8		22	
	10.1	10.6		21	
	7.6	11.5		20	
	11.8	14.3		19	
	14.2	13.6		18	
	10.8	9.3		17	
		10.9		16	
				15	
$\overline{X}$	12.33	11.42	2	14	13
Md	11.10	11.30		13	56
$\overline{X}$			9	12	0 — $\overline{X}$
Md			874	11	35 — Md
			851	10	69
			21	9	378
				8	
			6	7	8

The Greek letter Σ (capital sigma) stands for "the sum of" and is a symbol used frequently in statistics. Σx simply means "the sum of all the x's." Formulas with Σ may seem formidable, but, as we have just seen, Σ is simply shorthand for a relatively simple and familiar calculation. Σ is virtually the only mathematical symbol used in this book that is not common in basic algebra.

Table 2.1 presents some data on weights of flakes recovered from two bell-shaped storage pits in the same site. The back-to-back stem-and-leaf plot reveals that the flakes from Pit 1 bunch together between about 9 and 12 g, with one outlier at 28.6 g (to which we probably do not want to pay too much attention). The flakes from Pit 2 also bunch together, although the peak is more spread out and may even have a slight tendency to split into two. The center of the batch of flakes from Pit 2 would appear to be a little higher on the whole than for those from Pit 1. For the flakes from Pit 1, the mean (calculated by summing up all 12 weights and dividing the total by 12) is 12.33 g. For Pit 2, the mean (calculated by summing up all 13 weights and dividing the total by 13) is 11.42 g. Both means are indicated in their approximate positions along the stem in the stem-and-leaf plot.

We can be fairly happy with the mean as an index of the center for Pit 2; it does point to something like the center of the main bunch in the batch, as seen in the stem-and-leaf plot. When we look at Pit 1, however, we have cause for concern. The mean seems to be well above the center of the main bunch in the batch. It is "pulled up" quite strongly by the high outlier at 28.6 g, which has a major impact on the sum of the weights. Since we just observed that the Pit 1 batch has a somewhat lower level than the Pit 2 batch, it is alarming that the mean for Pit 1 is actually higher than the mean for Pit 2. A comparison of means for these two batches would suggest that flakes from Pit 1 tended to weigh more than those from Pit 2 – a conclusion exactly opposite to the one we arrived at by examining the stem-and-leaf plot. In this instance, the mean is not behaving very nicely. That is, it is not providing a useful index of the center of the Pit 1 batch for the purpose of comparing that batch to the Pit 2 batch. There are ho hard-and-fast rules for judging when the mean is behaving nicely enough to use as an index of center. It is finally a question of subjective judgment that requires careful exploration of batches with stem-and-leaf plots, real understanding of what we want an index of center to do, and practice.

THE MEDIAN

If the mean does not behave nicely because of the shape of a batch, the *median* may be a more useful index of center. The median is simply the middle number in the batch (if the batch contains an odd number of numbers) or halfway between the two middle numbers (if it contains an even number of numbers). The stem-and-leaf plot is useful for finding the median, because it makes it easy to count in from either the top or the bottom to the middle number. It is especially easy to do this if the leaves have been placed in numerical order on each line of the stem-and-leaf plot. The alternative to the stem-and-leaf plot, the histogram, cannot be used for finding the median because, while the histogram represents the overall shape of the batch, it does not contain the actual numbers.

To find the median weight of flakes from Pit 1, we first count the number of flakes. Since there are 12 (an even number), the median will be halfway between the middle two numbers. The middle two numbers will be the sixth and seventh, count-ing in from either the highest or lowest number. For example, counting leaves in the stem-and-leaf plot for Pit 1 from the bottom or lowest number, we have the first five numbers: 7.6, 9.1, 9.2, 10.1, and 10.5; then the sixth and seventh numbers: 10.8 and 11.4. Alternatively, counting leaves from the top or highest number, we have the first five numbers: 28.6, 14.2, 12.9, 11.8, and 11.7; then the sixth and seventh: 11.4 and 10.8, the same as before. Halfway between 10.8 and 11.4 is 11.1. So the median weight of flakes from Pit 1 is 11.10 g ($Md = 11.10$ g).

For Pit 2, there are 13 flakes, so the median will be the middle number, or the seventh in from either the highest or lowest. Counting leaves from the top gives us the first six numbers: 14.3, 14.1, 13.6, 13.5, 12.0, and 11.5; then the seventh: 11.3. Counting leaves from the bottom gives us the first six numbers: 7.8, 9.3, 9.7, 9.8,

10.6, 10.9; then the seventh: 11.3, exactly as before. Thus the median weight of flakes from Pit 2 is 11.30 g ($Md = 11.30$ g).

Medians for both batches are indicated on the stem-and-leaf plot in Table 2.1, and both indicate points that are visually more satisfying indications of the centers of the two batches. Comparing the levels of the two batches according to their medians also seems more reasonable than our attempt to use their means for this purpose. The median weight of flakes in Pit 2 is slightly higher than that for Pit 1, which is indeed the conclusion we came to based on observation of the general pattern of the stem-and-leaf plot.

OUTLIERS AND RESISTANCE

It might seem surprising that the mean and the median behave so differently in this example. After all, both are fairly widely used indexes of the level of a batch. And yet, comparing the two batches in this example by means and by medians gave opposite conclusions about which batch had a higher center. Clearly, it is the mean of the flakes from Pit 1 that seems strange. Its peculiarly high position is attributable entirely to the effect that the one high outlier (the flake that weighs 28.6 g) has on the calculations. While it pulls the mean up substantially, this outlier, in contrast, has no effect whatever on the median. If instead of weighing 28.6 g, this flake had weighed 12.5 g, the median flake weight for Pit 1 would not have changed at all. The heaviest flake is simply the first number that we count past to reach the middle of the batch, which remains in exactly the same place, irrespective of how high the highest value is. In fact, the median does not depend at all on the actual values of the numbers in either the upper half or the lower half of the batch. As long as there is no change that moves a number from the upper half to the lower half or vice versa, the median remains exactly the same.

This is one example of a general principle. The mean of a batch is strongly affected by any outliers that may be present. The median is entirely unaffected by them. In statistical jargon, the median is very *resistant*. The mean is not at all resistant.

ELIMINATING OUTLIERS

The mean has special properties that make it a particularly useful index of the center of a batch, but outliers can present a serious problem by making the mean a very inaccurate index. It would be nice to eliminate outliers if we could, and, as it turns out, often we can. In the first place, we should always examine out-liers carefully. Sometimes they indicate errors in data collection or recording. This possibility was already broached in Chapter 1, where it was suggested that the extraordinarily large post hole in the example in Fig. 1.1 might have been the result of an error in measurement or in data recording. Such an error could be corrected by reference to photographs and drawings of the excavation, thus eliminating the outlier.

Even if it turns out that an outlier is, indeed, a correct value, it still may be desirable to eliminate it. As a classic example of such a situation, consider the mail order clothing firm of L.L. Pea, Inc., specializing (of course) in the famous Pea coat. L.L. Pea employs ten shipping clerks, nine of whom are each paid $8.00 per hour while the tenth earns $52.00 per hour. The median wage in the L.L. Pea shipping room, then, is $8.00 per hour, while the mean wage is $12.40 per hour. Once again, the mean has been raised substantially by an outlier, while the median has been entirely unaffected. A careful check of payroll records reveals that it is, indeed, true that nine shipping clerks are paid $8.00 per hour while one earns $52.00 per hour. It also reveals, however, that the highly paid clerk is Edelbert Pea, nephew of L.L., the founder of the company, who spends most of his "working" hours in the company cafeteria anyway. If our interest is in the wages of shipping clerks, there is clearly no reason to include young Edelbert among our data. We are much better off simply to eliminate him as not truly a case of what we want to study and use the data for the other nine shipping clerks.

It is often sensible to eliminate outliers in just such a manner. If a good reason can be found aside from just the aberrant number in the data (as in the instance of Edelbert Pea), we can feel quite comfortable about eliminating outliers. In the example batch in Table 2.1 for Pit 1, perhaps we would note that the unusually heavy flake was of a very different form from all the rest or of a very different raw material. In this last case, we might reduce our batch to obsidian flakes, say, rather than all flakes, in order to eliminate a single very heavy chert flake. Even if such external reasons cannot be found to justify it, a distant outlier can be eliminated simply on the basis of its measurement. There are, however, other treatments that take care of outliers without making it seem that somehow we are fudging our data by leaving out cases we don't like.

THE TRIMMED MEAN

The *trimmed mean* systematically removes extreme values from both upper and lower ends of a batch in a balanced fashion. In considering the level of a batch, it is the central bunch of numbers that matters most. It is not uncommon for the highest and lowest numbers to straggle away from this bunch in an erratic manner, and it is important not to be confused by such unruly behavior on the part of a few numbers. The trimmed mean effectively avoids such confusion by simply eliminating some proportion of the highest and lowest numbers in the batch from consideration.

For example, we might calculate a 5% trimmed mean of the flake weights from Pit 1 in Table 2.1. For a 5% trimmed mean, we eliminate the highest 5% of the batch and the lowest 5% of the batch. There are 12 numbers in this batch, so we remove 5% of 12 numbers from each end. Since $0.05 \times 12 = 0.60$, and 0.60 rounds up to 1, we remove one number from the top and one number from the bottom. (In deciding how many numbers to remove for the trimmed mean we always round *up*.) In this case, then, we remove the highest number (28.6) and the lowest number (7.6) from the batch. After removing the highest and lowest numbers, we have a trimmed batch of ten numbers ($n_T = 10$). The trimmed mean is simply the ordinary mean of the remaining ten numbers, once the highest and lowest have been removed. For Pit 1 the 5% trimmed mean, $\overline{X}_T$, is the sum of the remaining numbers divided by n_T (that is, 10), or 11.17 g. For Pit 2, a 5% trimmed mean also requires eliminating a single number from each end of the batch ($0.05 \times 13 = 0.65$, which rounds up to 1). The total of the remaining numbers is divided by n_T (that is, 11), for $\overline{X}_T = 11.48$ g.

We can see that the trimmed mean, unlike the ordinary mean, is resistant to the effect of outliers. In this example, the 5% trimmed means are quite similar to the medians. They would lead us to conclude that flakes in Pit 2, in general, weigh slightly more than flakes in Pit 1, just as observation of the stem-and-leaf plot makes us know we should conclude.

In the 5% trimmed mean calculated above, 5% is the *trimming fraction*. The trimming fraction can be adjusted to fit the needs of a particular situation. Customarily, the trimming fraction is some multiple of 5% (5%, 10%, 15%, etc.). The most frequently used trimming fractions are probably 5% and 25%. The 25% trimmed mean is sometimes called the *midmean* because it is the mean of the middle half of the numbers (one-fourth of the numbers having been eliminated from the top of the batch and one-fourth from the bottom).

As one final example, a 25% trimmed mean of the flake weights from Pit 1 in Table 2.1 requires elimination of the three highest and the three lowest numbers ($0.25 \times 12 = 3$). The mean of the remaining six numbers is 11.05 g. For the flake weights from Pit 2, a 25% trimmed mean requires removal of four numbers from the top and bottom ($0.25 \times 13 = 3.25$, which rounds up to 4). The mean of the remaining five numbers is 11.26 g. Just as with the 5% trimmed mean, the undesirable effects of outliers have been avoided entirely; and the comparison of means shows that Pit 2 flakes are, in general, slightly heavier than Pit 1 flakes.

Statpacks

Any statistics package will determine the mean and median for a batch of numbers. Not very many, however, provide the trimmed mean as a defined option. What you are likely to have to do to get your statpack to calculate a trimmed mean is do the trimming yourself. You could simply omit the numbers to be trimmed when entering the data initially or you could delete those cases (or code them as missing data by whatever provision your statpack makes for handling missing data). Then your statpack can easily calculate the mean of the remaining numbers.

It is worth noting that the median could be thought of as the ultimate in trimmed means, the 50% trimmed mean. Removing the upper half of the batch and the lower half of the batch leaves nothing but the midpoint, or median.

WHICH INDEX TO USE

The median, the mean, and the trimmed mean are all numerical indexes of the center of a batch. The question thus arises, which one should we use? This question has no simple answer. Sometimes it is better to use the mean, sometimes the median, sometimes the trimmed mean. It depends on the characteristics of the batch in question and on what you intend to do with the numerical index of the center once you have it. The mean is the most familiar, and that is an advantage worth considering, since just about anyone feels comfortable if you tell them what the mean of a batch of numbers is. If the batch does not have outliers that make the mean a deceptive value, then it may well be the best choice. The median is slightly less familiar, but it is highly resistant, and so it is used fairly often for batches with outliers. The trimmed mean is considerably less familiar to most archaeologists, but it combines advantages of mean and median in some respects.

As we will see in later chapters, the mean has some special properties that make it highly useful in statistics. It is thus often tempting to use the mean, even when the batch has outliers that affect it. The trimmed mean can be put to work in at least some of the same ways the mean can, however, without interference from outliers. That is what makes the trimmed mean worth discussing, even though it is more complicated to calculate than either the mean or the median and less well known among archaeologists. The median, unfortunately, cannot be used in these special ways. Even though it is quite straightforward and useful for the initial task of comparing batches, then, the median will not be as important to us farther along in this book as the mean and the trimmed mean.

BATCHES WITH TWO CENTERS

Sometimes examination of a stem-and-leaf plot makes it clear that a batch contains two or more quite distinct bunches, as discussed in Chapter 1. We will call such batches *two-peaked* or *multi-peaked*. (The metaphor of the peak is derived from the histogram, where a bunch of numbers resembles a hill or peak, but it is easy enough to think of a stem-and-leaf plot in these terms as well.)

Table 2.2 provides the areas (in square meters) of structures excavated at the Black-Smith sites. The stem-and-leaf plot shows that these structures form two separate groups on the basis of their areas. There are large structures, mostly from about 15 to 21 m^2, and small structures, from about 3 to 7 m^2. It would make little sense to talk about the center of this batch because it clearly has two centers. If it makes little

**Table 2.2. Floor Areas of Structures at the
Black-Smith Sites**

Area (m^2)	Stem-and-leaf plot	
18.3	26	8
18.8	25	
16.7	24	
6.1	23	4
5.2	22	
21.2	21	2
19.8	20	07
4.2	19	128
18.3	18	33789
3.6	17	59
20.0	16	27
7.5	15	03
15.3	14	
26.8	13	6
5.4	12	
18.7	11	
6.2	10	
7.0	9	
20.7	8	
18.9	7	05
19.2	6	1277
6.7	5	244689
19.1	4	259
23.4	3	6
4.5		
16.2		
5.6		
17.5		
5.9		
6.7		
4.9		
17.9		
15.0		
13.6		
5.4		
5.8		

sense to talk about its center, then it makes even less sense to calculate a numerical index of its center. If we tried it, the results would be nonsense. The mean, for example, of the batch in Table 2.2 would be 12.95 m^2. This value falls in between the two distinct groups, characterizing no structures at all. At 15.15 m^2, the median would also fail to characterize the center of anything meaningful. We would thus never even calculate these two values.

The first thing to do if you see a two-peaked batch in a stem-and-leaf plot is separate it into two different batches – before calculating any indexes of center. This is not some mysterious rule that must be memorized. It is simply the only practice that makes sense to anyone who keeps firmly in mind what indexes of center are doing and how they behave. In a case like this, one must think that there are basically two different kinds of structures represented, perhaps houses and grain bins. Other information concerning these structures could be examined for evidence relevant to such a notion. In any event, before further quantitative analysis the batch must be broken into two batches, and the large structures treated separately from the small structures. We would make the break at about 10 or 11 m^2 in the middle of the large gap visible in the stem-and-leaf plot. The 16 small structures that are less than 10 m^2 have a mean area of 5.67 m^2 (and an almost identical median area of 5.70 m^2). The 20 large structures have a mean area of 18.77 m^2 (and, once again, an almost identical median area of 18.75 m^2). For both small structure areas and large structure areas, then, either the mean or the median would provide meaningful and useful indexes of the center. (Locate them along the stem in the stem-and-leaf plot, and you will see that they are indeed in the center of the main bunch of numbers for each sub-batch.) Breaking a two-peaked batch into two batches has made it possible to calculate numerical indexes of the centers of the two batches that make sense.

Batches like the one in Table 2.2 are often referred to loosely as *bimodal*, after the term *mode* which refers to the single most common category in a stem-and-leaf plot or histogram. Sometimes the mode is used as an index of the center of a batch. In Table 2.2, the mode would be at about 5 m^2, where six structures fall. This, clearly, is something like the center of the batch of small structures, but it won't do as an index of the center of the entire batch. There is a secondary mode at about 18 m^2, where five structures fall. This is something like the center of the batch of large structures. Only if exactly the same number of structures fell at 5 m^2 and at 18 m^2 would this batch truly have two modes. Strictly speaking, it has a mode and a secondary mode rather than two modes. Nevertheless, such multipeaked batches are often referred to as bimodal.

PRACTICE

1. Look back at the data on scraper lengths given in Tables 1.9 and 1.10. Calculate appropriate indexes of center to put a finer point on the comparison you have already made with a stem-and-leaf plot between Pine Ridge and the Willow Flats scraper lengths. Try out the mean, the median, and a trimmed mean (with whatever trimming fraction you think is most appropriate). Which index of center makes most sense for the comparison of scraper lengths between the two sites? Why? (Note that comparisons of levels must be based on the same index. You shouldn't compare the mean for one batch to the median for another.) Summarize the comparison of scraper lengths you have made between the two sites. That is, what has all this told you about scraper lengths at the two sites?

2. Using the data from Tables 1.9 and 1.10 once again, do the same for flint scrapers and chert scrapers, disregarding which site the scrapers came from. Try the mean, the median, and the trimmed mean again. Which index makes most sense for comparing the lengths of scrapers made of different raw materials? Why? How would you summarize all together the comparisons you have made between flint and chert scrapers and between the Willow Flats site and Pine Ridge Cave?

Chapter 3
The Spread or Dispersion of a Batch

Some batches of numbers are very tightly bunched together while others are much more spread out. This property is referred to in exploratory data analysis as *spread* (or in more traditional statistical terms as *dispersion*), and it is often an informative characteristic of a batch to which you should pay attention. Just as it is convenient to have a numerical index for the level or center of a batch, it is also convenient to have a numerical index for the spread, or dispersion, of a batch. Once again there are several different numerical indexes that behave differently and are thus used in different circumstances.

THE RANGE

The simplest index of the spread of a batch is its *range*. The range in statistics is exactly what it is in everyday conversation: the difference between the lowest number and the highest number in the batch. Table 3.1 presents the same example numbers we discussed in the previous chapter. The range for the weights of flakes recovered from Pit 1 is the difference between 28.6 and 7.6 g, or 21.0 g ($28.6\,\text{g} - 7.6\,\text{g} = 21.0\,\text{g}$). The range for the weights of flakes recovered from Pit 2 is the difference between 14.3 and 7.8 g, or 6.5 g ($14.3\,\text{g} - 7.8\,\text{g} = 6.5\,\text{g}$).

We notice immediately that the range suffers from the same problem that the mean suffers from: it is not at all resistant. In fact, it is even less resistant than the mean. Not only is it strongly affected by outliers, it may well depend entirely on outliers. Examination of the stem-and-leaf diagram reveals how misleading the range is in this instance. The two batches here have rather similar spreads, but we would probably say that the flake weights from Pit 2 are more spread out than those

R.D. Drennan, *Statistics for Archaeologists*, Interdisciplinary Contributions
to Archaeology, DOI 10.1007/978-1-4419-0413-3_3,
© Springer Science+Business Media, LLC 2004, 2009

Table 3.1. Weights of Flakes Recovered from Two Bell-Shaped Pits

	Flake weights (g)		Back-to-back stem-and-leaf plot		
	Pit 1	Pit 2	Pit 1		Pit 2
	9.2	11.3	6	28	
	12.9	9.8		27	
	11.4	14.1		26	
	9.1	13.5		25	
	28.6	9.7		24	
	10.5	12.0		23	
	11.7	7.8		22	
	10.1	10.6		21	
	7.6	11.5		20	
	11.8	14.3		19	
	14.2	13.6		18	
	10.8	9.3		17	
		10.9		16	
				15	
$\bar{X}$	12.33	11.42	2	14	13
Md	11.10	11.30		13	56
			9	12	0
Range	21.0	6.5	74	11	35
Midspread	3.7	3.7	851	10	69
			21	9	378
				8	
			6	7	8

of Pit 1 because the central bunch (which is always the most important part of the batch) is more dispersed along the stem. Nevertheless, the range for Pit 1 is much greater, entirely because of the one very high outlier in the Pit 1 batch. Although the range is simple to calculate and easily understood by everyone, it is likely to be very misleading unless all outliers can be removed. It is not much used as an index of spread.

THE MIDSPREAD OR INTERQUARTILE RANGE

The *midspread* is the range of the middle half of a batch. The highest 25% of the numbers and the lowest 25% of the numbers are thus disregarded. It could be thought of as a sort of trimmed range, thinking back to the trimmed mean discussed in Chapter 2.

In practice the midspread is found by locating the *quartiles* and subtracting the lower quartile from the upper quartile. The *upper quartile* is something like the median of the upper half of the batch and the *lower quartile* is something like the median of the lower half of the batch, although the rules used for finding the quartiles differ slightly from those used for finding the median. (In exploratory data analysis the quartiles are often called the *hinges*.) To find the quartiles, first divide the number of numbers in the batch by 4. If the result is a fraction, round it up to the next whole number. Then count in that many numbers from the highest number in the batch to arrive at the upper quartile and from the lowest number in the batch to arrive at the lower quartile.

For example, there are 12 flakes from Pit 1 for which weights are given in Table 3.1. We divide 12 by 4 and get 3. The upper quartile is the third number from the top of the stem-and-leaf, or 12.9 g. The lower quartile is the third number from the bottom of the stem-and-leaf, or 9.2 g. The midspread is then 12.9 g − 9.2 g = 3.7 g. For Pit 2, we have a batch of 13 weights; (13/4) = 3.25, which we round up to 4. The upper quartile is the fourth number from the top of the stem-and-leaf, or 13.5 g. The lower quartile is the fourth number from the bottom of the stem-and-leaf, or 9.8 g. The midspread is thus 13.5 g − 9.8 g = 3.7 g.

The midspread gives us better results for this example than the range, indicating that both batches are spread out to the same degree (a midspread of 3.7 g for both batches). This is at least closer to the mark than using a numerical index that shows the Pit 1 batch to be much more spread out than the Pit 2 batch.

The procedure for finding the midspread also reveals why it is sometimes called the *interquartile range* (at least by those who never use two syllables when five will do). The midspread is simply the range between the quartiles, and interquartile range is the traditional term for it. The midspread is used more in exploratory data analysis than in traditional statistics, and it works particularly well with the median to give us a quick indication of the level and spread of a batch.

THE VARIANCE AND STANDARD DEVIATION

The *variance* and the *standard deviation* are based on the mean. They are considerably more cumbersome to calculate than the range or the midspread, and they lack some of the immediately intuitive meaning that the range and midspread have. They have technical properties, however, that make them extraordinarily useful, and so they will be of considerable importance to many of the following chapters.

The basic concept on which the variance is based is that of difference from the mean. Clearly the vast majority of numbers in a batch are likely to be rather different from the mean of the batch. We can easily see how different any number in a batch is from the mean by subtracting the mean from it. The first two columns of Table 3.2 illustrate this procedure for all the numbers in the batch of weights of flakes from Pit 2 in Table 3.1. As is logical, the higher numbers in the batch have positive deviations from the mean (because they are *above* the mean), and the lower numbers have negative deviations from the mean (because they are *below* the mean). The numbers at the extreme ends of the batch, of course, deviate quite strongly from the mean in

Table 3.2. Calculating the Standard Deviation of Flake Weights from Pit 2 (Table 3.1)

x (g)	Deviations from mean $x - \overline{X}$	Squared deviations from mean $\left(x - \overline{X}\right)^2$
14.3	2.88	8.29
14.1	2.68	7.18
13.6	2.18	4.75
13.5	2.08	4.33
12.0	0.58	0.34
11.5	0.08	0.01
11.3	−0.12	0.01
10.9	−0.52	0.27
10.6	−0.82	0.67
9.8	−1.62	2.62
9.7	−1.72	2.96
9.3	−2.12	4.49
7.8	−3.62	13.10
$\overline{X} = 11.42$	$\sum (x - \overline{X}) = -0.06$	$\sum \left(x - \overline{X}\right)^2 = 49.02$ (sum of squares)

$$s^2 = \frac{\sum (x - \overline{X})^2}{n-1} = \frac{49.02}{12} = 4.09$$
$$s = \sqrt{s^2} = \sqrt{4.09} = 2.02$$

either positive or negative direction. The more spread out a batch is, the more strong deviations from the mean there are.

If we want to summarize these deviations numerically, it might occur to us to take the mean of the deviations. This won't do, however, because we can see that the deviations must always add up to 0; hence, their mean will always be 0. Indeed, a different way to think of the mean is to consider it a "balance point" that makes these deviations add up to 0. (You may notice that the second column of Table 3.2 actually adds up to −0.06 rather than 0. This is a consequence of rounding error, which commonly occurs. All the deviations are rounded off to two digits following the decimal point, and in this case by pure chance a little more rounding down has occurred than rounding up.)

What we are interested in, as an index of spread, is the set of deviations from the mean without their signs. We could simply drop the signs and add up the absolute values of the deviations, but it turns out to be preferable to get rid of the signs by squaring the deviations from the mean. (The squares of the deviations from the mean are, of course, all positive, as squares must all be.) This calculation is shown in the third column of Table 3.2. It is this third column that we sum up. This sum is sometimes referred to as the sum of the squared deviations from the mean or simply the *sum of squares*.

This sum of squares will, other things being equal, be larger for a larger batch of numbers than for a smaller batch because a larger batch has more deviations to add up. To arrive at an index that is not affected by the size of the batch but only by its spread, what we need is something like the average squared deviation from the mean. Instead of dividing the sum of squares by the number of numbers in the batch, however, we divide it by one less than the number of numbers in the batch. We do this for purely technical reasons to make the result more useful in future chapters where we take batches of numbers to be samples from larger populations. The equation for the variance, then, is

$$s^2 = \frac{\sum (x - \overline{X})^2}{n - 1}$$

where s^2 is the variance of x, $\overline{X}$ is the mean of x, and n is the number of numbers in the batch of x.

Table 3.2 provides an example of the calculations that correspond to this equation. The variance has a rather arbitrary character compared to the range or the midspread. The value of the variance is not as easy to relate intuitively to the values in the batch as was the case with the range or midspread. We can at least remove the confusing effect of squaring the deviations by taking the square root of the variance. The result is s, the standard deviation:

$$s = \sqrt{s^2} = \sqrt{\frac{\sum (x - \overline{X})^2}{n - 1}}$$

The standard deviation, unlike the variance, is at least expressed in the same units as the original batch. Thus it is appropriate to think of the standard deviation of the weights of flakes from Pit 2 as not just 2.02, but 2.02 g. If we relate the standard deviation to the stem-and-leaf plot in Table 3.1, we see that the standard deviation delineates the portion of the stem within which most of the flake weights fall. That is, most weights are within 2.02 g above or below the mean of 11.42 g, which is to say, most of the weights are between 9.40 (11.42 g − 2.02 g = 9.40 g) and 13.44 g (11.42 g + 2.02 g = 13.44 g). These two numbers (9.40 and 13.44 g) provide an approximation of the limits of the main bunch of numbers. That is what it means to say that most of the flake weights are within one standard deviation of the mean. Only a few fall farther than one standard deviation from the mean, that is, farther than 2.02 g from the mean. We can (and will) specify much more about this way of using the standard deviation in later chapters. For the moment, suffice it to say that the standard deviation often provides just this kind of indication about the spread of a batch.

The standard deviation does not behave so satisfactorily for the flake weights from Pit 1. Table 3.3 shows the calculation of the standard deviation for this batch. When we first compared these two batches of numbers (the weights of flakes from Pits 1 and 2) on the basis of the stem-and-leaf plots in Table 2.1, we noted that the flake weights from Pit 1 were (except for the high outlier) more closely bunched up than those from Pit 2. The variance and the standard deviation for flake weights from Pit 1, however, are much larger than those for Pit 2, indicating a much larger

Table 3.3. Calculating the Standard Deviation of Flake Weights from Pit 1 (Table 3.1)

x (g)	Deviations from mean $x - \overline{X}$	Squared deviations from mean $\left(x - \overline{X}\right)^2$
28.6	16.27	264.71
14.2	1.87	3.50
12.9	0.57	0.32
11.8	−0.53	0.28
11.7	−0.63	0.40
11.4	−0.93	0.86
10.8	−1.53	2.34
10.5	−1.83	3.35
10.1	−2.23	4.97
9.2	−3.13	9.80
9.1	−3.23	10.43
7.6	−4.73	22.37
$\overline{X} = 12.33$	$\Sigma(x - \overline{X}) = -0.06$	$\Sigma\left(x - \overline{X}\right)^2 = 323.33$
		(sum of squares)

$$s^2 = \frac{\Sigma(x - \overline{X})^2}{n - 1} = \frac{323.33}{11} = 29.39$$

$$s = \sqrt{s^2} = \sqrt{29.39} = 5.42$$

spread for the flakes from Pit 1 – exactly opposite the conclusion the stem-and-leaf plot clearly indicates.

Table 3.3 shows very clearly why the variance and the standard deviation are so large for Pit 1: the value for the one heaviest flake deviates very strongly from the mean. That one flake is alone responsible for such a high sum of squares and thus for such a high variance and standard deviation. Clearly, like the mean, the variance and the standard deviation are not at all resistant to the effects of outliers. Using the variance or the standard deviation as a numerical index of the spread of a batch, then, is not a good idea at all if the batch has outliers.

Table 3.3 also provides a convenient illustration of why the mean lacks resistance along the lines of observations made in Chapter 2. Think of the mean as the balance point of a see-saw. The high outlier is like a person far out at one end of the see-saw. In order to make the see-saw balance, the mean must be moved substantially toward that end so that most of the numbers are on the other side. In that position it is far off to one side of the center of the main bunch of numbers. It was precisely this undesirable effect that we complained about in Chapter 2.

THE TRIMMED STANDARD DEVIATION

The basic idea of the trimmed standard deviation is exactly like that of the trimmed mean: outliers are excluded from the sample so that they will not have an undue

**Table 3.4. Calculating the 5% Trimmed Standard Deviation
of Flake Weights from Pit 1 (Table 3.1)**

Original batch x (g)	Winsorized batch x_W (g)	Deviations from mean $x_W - \overline{X}_W$	Squared deviations from mean $(x_W - \overline{X}_W)^2$
28.6	14.2	2.95	8.70
14.2	14.2	2.95	8.70
12.9	12.9	1.65	2.72
11.8	11.8	0.55	0.30
11.7	11.7	0.45	0.20
11.4	11.4	0.15	0.02
10.8	10.8	−0.45	0.20
10.5	10.5	−0.75	0.56
10.1	10.1	−1.15	1.32
9.2	9.2	−2.05	4.20
9.1	9.1	−2.15	4.62
7.6	9.1	−2.15	4.62

$$\overline{X}_W = 11.25 \quad \Sigma(x_W - \overline{X}_W) = 0.00 \quad \Sigma(x_W - \overline{X}_W)^2 = 36.16$$
$$\text{(sum of squares)}$$

$$s_W^2 = \frac{\Sigma(x_W - \overline{X}_W)^2}{n-1} = \frac{36.16}{11} = 3.29$$

$$s_T = \sqrt{\frac{(n-1)s_W^2}{n_T - 1}} = \sqrt{\frac{(12-1)3.29}{(10-1)}} = 2.01$$

effect on the result. Calculation of the trimmed standard deviation, however, becomes more involved. Instead of simply reducing the size of the batch by trimming off numbers at the top and bottom, we must maintain the size of the batch by replacing trimmed numbers with the numbers next in line for trimming. Table 3.4 shows this process for calculating a 5% trimmed standard deviation of the batch of flake weights from Pit 1. When, in Chapter 2, we calculated the 5% trimmed mean of this same batch, we trimmed the single highest and lowest number from the batch. This time, we replace the highest number with the next highest number (the highest number that remained in the batch after trimming). Thus 28.6 g becomes 14.2 g. Similarly, we replace the lowest number with the next lowest number (the lowest number that remained in the batch after trimming). Thus 7.6 g becomes 9.1 g.

The new batch that results is a *Winsorized batch*. The Winsorized variance is calculated simply as the ordinary variance of this Winsorized batch. Note, though, that the mean involved in calculating the Winsorized variance is the mean of the Winsorized batch (which is not the same as the trimmed mean) and that the trimmed standard deviation is *not* simply the square root of the variance of the Winsorized batch. The trimmed standard deviation is derived from the Winsorized variance by the following equation:

$$s_T = \sqrt{\frac{(n-1)s_W^2}{n_T - 1}}$$

Statpacks

Midspreads and standard deviations are pretty common fare in statpacks, and statpacks are truly helpful here because calculating a standard deviation with a calculator is time consuming (unless your calculator has a special key for doing it automatically). Trimmed standard deviations, however, are much less often provided for in statpacks. Just as in calculating a trimmed mean with your statpack, you are likely to have to adjust the batch yourself first. In this case instead of replacing extreme values with missing data, you replace extreme values with the adjacent nonextreme value in the data. Once this modification has been made, the batch has been Winsorized, and the variance your statpack calculates on these numbers is the Winsorized variance, which you can convert into the trimmed standard deviation with your calculator, as illustrated in Table 3.4. *Be sure not to forget this last step!*

---

where s_T is the trimmed standard deviation, n is the number of numbers in the untrimmed batch, s_W^2 is the variance of the Winsorized batch, and n_T is the number of numbers in the trimmed batch.

Table 3.4 shows the full calculation of the trimmed standard deviation for the flake weights from Pit 1. Comparison of the calculation columns for Tables 3.3 and 3.4 shows quite clearly how the trimmed standard deviation avoids the overwhelming effect of outliers.

Just as the trimmed mean can be calculated for various trimming fractions, so can the trimmed standard deviation. In Chapter 2 we calculated a 25% trimmed mean of the flake weights from Pit 1 by trimming the three highest and the three lowest numbers from the batch. Calculation of the 25% trimmed standard deviation would begin with the creation of a Winsorized batch of 12 numbers in which the three highest numbers were replaced with the fourth highest and the three lowest numbers were replaced with the fourth lowest. From there on the calculation of the variance of the Winsorized batch and the trimmed standard deviation follow exactly the same path we have just taken for the 5% trimmed standard deviation. When a trimmed mean and standard deviation are used, the trimming fraction should always be specified.

WHICH INDEX TO USE

The range, the midspread, the standard deviation, and the trimmed standard deviation are all numerical indexes of the spread of a batch. Just as we asked when to use which index of the center of the batch, we must ask when to use which index of spread. The answer parallels that given in Chapter 2. The range is very widely understood but so badly affected by outliers that it is not often of much use. The midspread has been emphasized in exploratory data analysis. It is not as familiar as it should be to archaeologists, but it is easy to find and of wide utility for basic descriptive purposes. Its resistance to the effects of outliers makes it particularly attractive. The standard deviation is quite widely familiar (at least the term is, whether or not many archaeologists are really at home with the concept or not). Its statistical properties, like those of the mean, will serve us well in the rest of this book. It is of such importance that we will spend some effort on techniques to overcome its poor resistance to the effects of outliers. Some of these techniques are based on the trimmed standard deviation. Indexes of center and spread work together in pairs: the median with the midspread, the mean with the standard deviation, or the trimmed mean with the trimmed standard deviation (both with the same trimming fraction). Using the median together with the standard deviation, for example, is like wearing one white sock and one brown sock – only worse.

Table 3.5. Areas of Bronze Age Sites
Near Nanxiong

Site area (ha)	
Early Bronze Age	Late Bronze Age
1.8	10.4
1.0	5.9
1.9	12.8
0.6	4.6
2.3	7.8
1.2	4.1
0.8	2.6
4.2	8.4
1.5	5.2
2.6	4.5
2.1	4.1
1.7	4.0
2.3	11.2
2.4	6.7
0.6	5.8
2.9	3.9
2.0	9.2
2.2	5.6
1.9	5.4
1.1	4.8
2.6	4.2
2.2	3.0
1.7	6.1
1.1	5.1
	6.3
	12.3
	3.9

PRACTICE

Imagine you have conducted a regional survey of a small valley north of Nanxiong and have carefully measured the areas of the surface scatters that indicate the Bronze Age sites you encountered. The areas (in hectares) are given in Table 3.5.

1. Begin to explore these two batches of numbers with a back-to-back stem-and-leaf plot.
2. Continue your exploration by calculating the median, the mean, and the 10% trimmed mean for each batch and then the index of spread that corresponds to each of these indexes of level. Which pair of indexes makes most sense to use here? Why?
3. Based on the stem-and-leaf plots and the indexes of level and spread, what observations would you make about changes in site size from Early Bronze Age to Late Bronze Age near Nanxiong?

Chapter 4
Comparing Batches

We have already compared batches with back-to-back stem-and-leaf plots, but there are quicker and more effective tools for graphically comparing batches. The numerical indexes of the center and spread of a batch that we have discussed in the last two chapters provide the basis for such tools. A standard way of plotting some of these indexes in exploratory data analysis is called the *box-and-dot* plot (or the *box-and-whisker* plot). The box-and-dot plot could, in theory, be based on any of the indexes of center and spread, but in practice the median and the midspread are used. This is such standard practice that a box-and-dot plot is automatically taken to represent the median and midspread, and this convention should not be violated.

THE BOX-AND-DOT PLOT

Construction of a box-and-dot plot begins in exactly the same way as construction of a stem-and-leaf plot: with the establishment of a scale along which the numbers in the batch will lie. Figure 4.1 presents a stem-and-leaf plot of post hole diameters from the Smith site (taken from Tables 1.7 and 1.8). To the right the stem is converted into a scale for drawing a box-and-dot plot. A horizontal line is placed next to 17.2 cm on this scale to represent the median. Two more lines at 18.8 and 15.7 cm represent the upper and lower quartiles. These three lines are framed with two vertical lines to form a box with a line across it near its center. This box graphically represents the midspread, that is, the central half of the numbers – those that fall between the two quartiles. The box provides a clear, clean picture of the most important central bunch of numbers in the batch, one that is more quickly perceived than the stem-and-leaf plot.

R.D. Drennan, *Statistics for Archaeologists*, Interdisciplinary Contributions
to Archaeology, DOI 10.1007/978-1-4419-0413-3_4,
© Springer Science+Business Media, LLC 2004, 2009

20	5
19	4
18	3689
17	29
16	48
15	379
14	3
13	
12	
11	
10	
9	
8	4

Median = 17.2 cm
Upper Quartile = 18.8 cm
Lower Quartile = 15.7 cm
Midspread = 3.1 cm

Figure 4.1. Box-and-dot plot of post hole diameters (in cm) from the Smith site.

We can include more detail in the box-and-dot plot, and at the same time provide more precise definition of other important features of a batch. We have, for example, already discussed outliers, numbers that fall far outside the central bunch and are generally a nuisance as far as several otherwise very useful numerical indexes of center and spread are concerned. It is often quite helpful to simply eliminate outliers, but we are frequently confronted with borderline cases – numbers that lie beyond the central bunch but not so far outside it as to make us certain that they do not belong with the batch.

The box-and-dot plot provides a graphical approach to identifying outliers consistently and signaling their presence, by suggesting a rule of thumb to distinguish between the central bunch of numbers and outliers. According to this rule of thumb, an outlier is any number that lies more than one and a half times the length of the box beyond either end of the box. We can think of this in purely graphical terms. We could measure the box in a box-and-dot plot as drawn on paper. If the box is 1 in. long, then we would say that any number falling more than 1.5 in. above the top of the box or below the bottom of the box is an outlier. Further, any number falling more than twice this far from the end of the box is a *far outlier*. These distances are indicated with lines in Fig. 4.1.

The same result can be achieved mathematically. Since the length of the box is the midspread, the distance that defines outliers is 1.5 times the midspread ($1.5 \times 3.1 = 4.65$ cm in the example in Fig. 4.1). Since the top of the box represents the upper quartile, the position of the defining line for high outliers on the number scale is the upper quartile plus 1.5 times the midspread ($18.8 + 4.65 = 23.45$ cm in the example

---

Rules of Thumb

Practical statistics is filled with rules of thumb that are efforts to patch the gaps between nice neat principles and much messier real life. Outliers create just such a gap. We considered the case of Edelbert Pea (the boss's nephew) as an example of an outlier. He was easy to identify as an outlier because he made $52.00 an hour while all the other shipping clerks made $8.00 an hour. But what if Edelbert made only $8.50 an hour? And what if he had worked at L.L. Pea for three years as a shipping clerk, while all the other shipping clerks who made $8.00 an hour had less than six months' experience? He no longer seems such an outlier. In fact, he begins to sound like a good example of exactly the kind of variation we would like to include in our study of shipping clerks' wages. But where would we draw the line? If Edelbert made $12.00 an hour would he be an outlier? If he made $20.00? In seeking to draw a line at the point where Edelbert's wages make him an outlier, we're basically trying to do the impossible. It makes no sense to pretend to be able to say, for example, that if he made as much as $14.73 an hour he would not be an outlier, but if he made $14.74 he would be. The judgment is just much fuzzier than that. On the other hand, if we're going to analyze shipping clerks' wages we have to either include Edelbert or exclude him. There is no middle ground. "Maybe" simply does not lead to any course of action we can pursue. It is in precisely such situations that statisticians make up rules of thumb – to provide systematic guidance where the best answer is "maybe" but the only useful answers are "yes" and "no."

Saying that a number is an outlier if it falls more than 1.5 midspreads outside either quartile of its batch is a rule of thumb. It provides us with a systematic way of identifying outliers in a batch according to a clearcut rule. But it would be hard to justify choosing exactly 1.5 midspreads rather than 1.6 or 1.4 because the choice, finally, is somewhat arbitrary. Indeed, there is some variation from one statistics book (or one computer program) to the next in the exact rule of thumb used for identifying outliers. The same is true of the other rules of thumb that we will discuss as we go on through this book.

in Fig. 4.1). Since the bottom of the box represents the lower quartile, the position of the defining line for low outliers is the lower quartile minus 1.5 times the midspread ($15.7 - 4.65 = 11.05$ cm in the example in Fig. 4.1).

In the same way, the positions of the defining lines for far outliers can be established mathematically. The line defining far high outliers is twice as far above the upper quartile as the line defining outliers. That is to say, instead of 1.5 times the midspread beyond the quartiles, the far outlier defining line falls at three times the midspread beyond the quartiles ($18.8 + 9.3 = 28.1$ cm and $15.7 - 9.3 = 6.4$ cm in Fig. 4.1).

Thus the areas above and below the box in the box-and-dot plot are each divided into three zones. Numbers that fall in the nearest zone above or below the box are called *adjacent values*. These numbers are outside the central half of the batch but are still considered part of the main bunch of numbers. In the next zone away from the median come outliers, and in the farthest zone are far outliers. Ordinarily these zones are not indicated by lines the way they are in Fig. 4.1. Instead, they are distinguished by different symbols representing the numbers that fall in them. The highest and lowest adjacent values are indicated with X's, as shown in Fig. 4.1. These X's, then, represent the extremes of the main bunch of numbers (excluding all outliers). Outliers are all indicated individually on the plot as hollow dots, and far outliers are all indicated individually as solid dots. The batch represented in Fig. 4.1 has only one outlier (8.4 cm) and no far outliers, so there is a single hollow dot and no solid dots. These conventions about X's, hollow dots, and solid dots stand for the labels and lines drawn to the right of Fig. 4.1, so such labels and defining lines do not generally appear when box-and-dot plots are drawn. As is the case with rules of thumb, the exact conventions used to indicate outliers and far outliers in box-and-dot plots vary from one book or program to the next.

The box-and-dot plot makes it easy to compare several batches. In Chapter 1, we compared the batch used for the example in Fig. 4.1 to another batch of post hole diameters with a back-to-back stem-and-leaf plot (Table 1.7). Figure 4.2 compares the same two batches with two box-and-dot plots instead. The box-and-dot plot for post hole diameters at the Smith site is exactly the same as in Fig. 4.1 (except that it is now on a longer scale). The box-and-dot plot for post hole diameters at the Black site is made in exactly the same manner, but using the numbers listed in Table 1.1 for the Black site. The one extremely large post hole qualifies not only as an outlier, but as a far outlier, since it lies more than three times the length of the box from the box's upper end. It is thus shown as a solid dot.

When we look at the box-and-dot plots in Fig. 4.2, we quickly reach the same conclusion we reached looking at the back-to-back stem-and-leaf plot of these same numbers in Table 1.7. At each site there is a post hole that does not seem to represent the same kind of phenomenon as the rest of the post holes – an extremely large post hole at the Black site and an extremely small post hole at the Smith site. In general, post holes at the Smith site are larger than post holes at the Black site by a margin of 5 or 6 cm. The box-and-dot plot shows us these patterns even more clearly than the back-to-back stem-and-leaf plot because the box-and-dot plot is a simpler, more quickly perceived way of representing the basic features of each batch. The

Figure 4.2. Box-and-dot plot comparing post hole diameters (in cm) at the Black–Smith sites.

box-and-dot plot can also be extended easily to the comparison of a larger number of separate batches simply by adding additional boxes and dots to the same scale. The back-to-back stem-and-leaf plot cannot be extended very conveniently to the comparison of more than two batches.

REMOVING THE LEVEL

When we compare two or more batches of numbers, as in Fig. 4.2, probably the most noticeable characteristic of each batch is its level or center. If we want to compare other features of the batches, it is convenient to remove the conspicuous effect of their differing levels. We do this by reducing the levels of both batches to zero.

Figure 4.3 shows this graphically. We have simply slid both box-and-dot plots down the scale so that the center of each (that is, the median) lines up with the zero point on the scale. The same result can also be achieved mathematically by subtracting the median of a batch from each number in the batch. For example, we take all the post hole diameters from the Smith site in Table 1.6, and subtract the median of Smith site post hole diameters (17.2 cm) from each one, as shown in Table 4.1. The result is a new set of numbers that represent how much each post hole is larger or smaller than the median size. Post holes whose diameter is larger than the median are represented by positive numbers and post holes whose diameter is smaller than the median are represented by negative numbers. We could arrive at the Smith site box-and-dot plot with the level removed (in Fig. 4.3) by making a box-and-dot plot of this new batch of numbers. The result would be exactly the same as graphically sliding the box-and-dot plot made previously down the scale until its median arrived at zero. (If you do not immediately see why this is so, the best way to understand is to try it out for yourself.)

Having removed the levels from these two batches of numbers we can no longer compare them in regard to level. The process of removing the level is to artificially set the center of both batches at zero. With the conspicuous effect of differences in level removed, however, we very quickly notice that the two batches differ in regard to spread. Disregarding the outliers and far outliers, we see that the adjacent values in both batches are similarly spread out on the number scale. The most central bunch of numbers, however (the middle half as represented by the box), is more spread out for the Smith site post holes than for the Black site post holes. This difference was certainly visible in the previous box-and-dot plot (Fig. 4.2), but it is considerably more conspicuous now that the two boxes have been lined up at their middles by removing the levels.

REMOVING THE SPREAD

Just as we removed the level from a batch by reducing its center to zero, we can remove the spread from a batch by reducing its spread to one. This must be accomplished mathematically; it cannot be done graphically, as in the case of removing the level by sliding the box down the number scale. Once the level has been removed mathematically, however, by subtracting the median, we remove the spread by dividing by the midspread. Table 4.2 continues where the calculations in Table 4.1 left off with the Smith site post hole diameters. The first number in the batch, for example, in Table 4.1 represents a post hole 20.5 cm in diameter. When the level is removed

Figure 4.3. Box-and-dot plots of post hole diameters (in cm) at the Black–Smith sites with levels removed.

from this number, we see that this post hole diameter is 3.3 cm larger than the median. Continuing with the calculation in Table 4.2, we see that this 3.3 cm divided by 3.1 cm (the midspread) is 1.06. This result, 1.06, means that the post hole diameter in question is above the median by an amount equal to a little bit more than one midspread. In the box-and-dot plot (Fig. 4.1), this post hole would lie about the

Table 4.1. Removing the Level from Smith Site Post hole Diameters by Subtracting the Median (17.2 cm)

20.5 cm	−	17.2 cm	=	3.3 cm
17.2 cm	−	17.2 cm	=	0.0 cm
15.3 cm	−	17.2 cm	=	−1.9 cm
15.9 cm	−	17.2 cm	=	−1.3 cm
18.3 cm	−	17.2 cm	=	1.1 cm
17.9 cm	−	17.2 cm	=	0.7 cm
18.6 cm	−	17.2 cm	=	1.4 cm
14.3 cm	−	17.2 cm	=	−2.9 cm
19.4 cm	−	17.2 cm	=	2.2 cm
16.4 cm	−	17.2 cm	=	−0.8 cm
18.8 cm	−	17.2 cm	=	1.6 cm
15.7 cm	−	17.2 cm	=	−1.5 cm
18.9 cm	−	17.2 cm	=	1.7 cm
16.8 cm	−	17.2 cm	=	−0.4 cm
8.4 cm	−	17.2 cm	=	−8.8 cm

length of the box above the median (that is above the center line of the box). Since this post hole provides the highest adjacent value, it is, in fact, located in Fig. 4.1 as the X above the box, and the center of this X does, indeed, lie about the length of the box above the box's center line.

Removing the level and spread from both batches of numbers, the post hole diameters from the Black site and from the Smith site, and making yet another box-and-dot plot of the result, gives us Fig. 4.4. The centers of both batches are still at zero, but now the boxes representing the middle half of each batch are the same length. That length, of course, is one, since the box length always represents the midspread, and removing the level and spread has the effect of setting the center at zero and the spread at one. We thus cannot use Fig. 4.4 to compare the batches in regard to either level or spread. The feature that becomes most conspicuous at this point is shape, especially symmetry. Clearly, the post hole diameters from the Black site tend to spread out downward from the median more than upward. Remember that one-quarter of the numbers in the batch fall between the median and the top of the box and one-quarter fall between the median and the bottom of the box. The one-quarter of the numbers immediately below the median at the Black site are clearly more spread out than the one-quarter immediately above the median, which clump closer to the median. The Smith site post hole diameters, on the other hand, have a more symmetrical distribution, although its middle half spreads upward a little more than it does downward. We will discuss shape and symmetry in more detail in the next chapter.

It is worth noting that there is an easier way to draw a box-and-dot plot with the level and spread removed. We have just subtracted the median from all the numbers in the batch and divided all the resulting numbers by the midspread to arrive at a new batch. In this batch we found the median, upper and lower quartiles, outliers, etc. so

Table 4.2. Removing the Spread from Smith Site Post hole Diameters by Dividing by the Midspread (3.1 cm) after the Level Has Been Removed (Compare to Table 4.1)

3.3 cm	/	3.1 cm	=	1.06
0.0 cm	/	3.1 cm	=	0.00
−1.9 cm	/	3.1 cm	=	−0.61
−1.3 cm	/	3.1 cm	=	−0.42
1.1 cm	/	3.1 cm	=	0.35
0.7 cm	/	3.1 cm	=	0.23
1.4 cm	/	3.1 cm	=	0.45
−2.9 cm	/	3.1 cm	=	−0.94
2.2 cm	/	3.1 cm	=	0.71
−0.8 cm	/	3.1 cm	=	−0.26
1.6 cm	/	3.1 cm	=	0.52
−1.5 cm	/	3.1 cm	=	−0.48
1.7 cm	/	3.1 cm	=	0.55
−0.4 cm	/	3.1 cm	=	−0.13
−8.8 cm	/	3.1 cm	=	−2.84

as to draw a new box-and-dot plot from scratch. We could simply have applied this treatment to each of the five numbers required to define the box-and-dot plot (the median, the upper and lower quartiles, and the upper and lower extreme adjacent values). These five values, with the level and spread removed, produce the same box-and-dot plot as the same five values determined afresh from a complete new batch with the level and spread removed from all the numbers. To finish the graph requires only subtracting the median from each outlier and dividing the result by the midspread so as to locate outliers on the new number scale.

UNUSUALNESS

This new number scale is a very interesting scale. It is no longer a scale of centimeters as the previous number scales have been, but rather, in effect, a scale of unusualness. It locates each number in the batch according to just how central or how peripheral that number is in terms of the batch to which it belongs. Unusualness is not an inherent property of a thing but rather a statement of how a thing relates to the group of which it is a member. If a thing falls well within the central bunch of things in its group, then it is not very unusual. If a thing falls in a more peripheral position, relative to the central bunch of things in its group, then it is more unusual. In a group of professional basketball all-stars, a person 6'-6" tall is not very unusual. In a group of university professors, however, a person 6'-6" tall is very unusual. Removing the level and spread from a batch of numbers gives us a scale along which we can express unusualness in a standard and systematic manner. For this reason the traditional statistical term for this procedure is *standardizing*.

Figure 4.4. Stem-and-leaf and box-and-dot plots of post hole diameters at the Black–Smith sites with levels and spreads removed.

The number scale in Fig. 4.4 expresses how far each number in each batch departs from the median for that batch in terms of the midspread for that batch. The first number in Table 4.1, for example, is 20.5 cm, which represents a post hole 3.3 cm larger than the median diameter for the Smith site. This post hole measurement becomes 1.06 in the standardized batch (Table 4.2), meaning that its diameter is slightly more than 1 midspread above the median. The second measurement in Table 4.1, 17.2 cm *is* the median. Thus its difference from the median is 0.0 cm or 0.00 midspreads. The third measurement, 15.3 cm, falls 1.9 cm *below* the median and becomes -0.61 in the standardized batch. It is thus 0.61 midspreads *below* the median value. The first post hole, then (20.5 cm in diameter), is more unusual than the third post hole, because it falls farther out toward the periphery of the batch.

Statpacks

As in the case of stem-and-leaf plots, there are many statpacks that draw box-and-dot plots. Their conventions for indicating outliers may vary from those used in this book, but as long as you know what they are, that should not pose a problem. Some programs draw box-and-dot plots vertically, the way they are drawn in this book, although sometimes the lower numbers are higher on the screen, and the higher numbers, lower on the screen in contrast to the figures here. Some programs draw the plots horizontally. None of this makes any difference, of course, to the interpretation of the plots. Usually such programs automatically choose a scale for the plots, releasing your time and energy for other more important tasks. If your program does not automatically produce box-and-dot plots of several batches all at the same scale for comparing several batches, you may need to look up how to take active control of determining the scale to be used. Clearly, box-and-dot plots of different batches cannot be compared to each other unless they are drawn to the same scale.

The easiest way to make box-and-dot plots with the level or level and spread removed, of course, is also with a statpack. Usually the procedure you need to follow is to *transform* the numbers in the original batch by subtracting the median from each (and dividing the result by the midspread if you want to remove the level and the spread) to create a new batch (or *variable*). Almost all statpacks make it easy to do such a thing. Then you can make a box-and-dot plot of the new batch.

This standardized number scale permits comparisons of unusualness from one batch to another. For example, the first post hole in Table 4.3, with a diameter of 9.7 cm, is 1.4 cm smaller than the median diameter at the Black site. The 15.7-cm post hole at the Smith site (fourth from the bottom in Tables 4.1 and 4.2) is 1.5 cm smaller than the median diameter in its batch. It might seem that this latter post hole is more unusual since it is farther from the center of its batch in centimeters. It is, however, in a batch that is just more spread out in general. At the Black site a post hole 1.4 cm smaller than the median lies 0.67 midspreads away from the median. At the Smith site a post hole 1.5 cm smaller than the median lies only 0.48 midspreads away from the median. The post hole with a diameter of 9.7 cm is thus more unusual for the Black site than is the post hole with a diameter of 15.7 cm for the Smith site.

Perhaps the context in which we most frequently encounter such unusualness scales is in standardized testing. Elementary school test results are often expressed in terms of how far above or below average a score is for a particular grade level. The *percentiles* in which college entrance examinations are commonly expressed also provide such information. A student who scores in the 75th percentile knows that about 75% of those taking the test had lower scores, while about 25% had higher scores. If the batch in question is symmetrical, the 75th percentile is equivalent to a score of about 0.5 on the unusualness scale we have been discussing. This is

Table 4.3. Removing the Level and Spread from Black Site Post hole Diameters by Subtracting the Median (11.1 cm) and Dividing by the Midspread (2.1 cm)

(9.7 cm – 11.1 cm)	/	2.1 cm	= −0.67
(9.2 cm – 11.1 cm)	/	2.1 cm	= −0.90
(12.9 cm – 11.1 cm)	/	2.1 cm	= 0.86
(11.4 cm – 11.1 cm)	/	2.1 cm	= 0.14
(9.1 cm – 11.1 cm)	/	2.1 cm	= −0.95
(44.6 cm – 11.1 cm)	/	2.1 cm	= 15.95
(10.5 cm – 11.1 cm)	/	2.1 cm	= −0.29
(11.7 cm – 11.1 cm)	/	2.1 cm	= 0.29
(11.1 cm – 11.1 cm)	/	2.1 cm	= 0.00
(7.6 cm – 11.1 cm)	/	2.1 cm	= −1.67
(11.8 cm – 11.1 cm)	/	2.1 cm	= 0.33
(14.2 cm – 11.1 cm)	/	2.1 cm	= 1.48
(10.8 cm – 11.1 cm)	/	2.1 cm	= −0.14

because when a batch is standardized by subtracting the median and dividing by the midspread, a score of 0.5 means above the median by half a midspread. A number that is half a midspread above the median is, of course, the upper quartile (at least in a symmetrical batch). And the upper quartile is the number above which lie 25% of the numbers in the batch.

STANDARDIZING BASED ON THE MEAN AND STANDARD DEVIATION

Expressing the unusualness of a number in terms of its centrality or peripheralness in its own batch is a critically important concept in statistics. Much of the remainder of this book is built on this concept of unusualness. In this chapter we have focused on removing the level and spread using the median and midspread as the numerical indexes of level and spread. We have done this because the box-and-dot plot based on the median and midspread provides a particularly easy graphical illustration of the procedure and its implications. It is more common and ultimately much more useful to use the mean and standard deviation, however, because these indexes have some especially attractive mathematical properties. The basic principles and the calculations are an exact parallel to what we have just discussed. To standardize a batch using the mean and standard deviation, you subtract the mean of the batch from every number in the batch and divide the result by the standard deviation of the batch. The resulting batch is often referred to as a batch of *standard scores* or *z scores*. The *z* scores tell how many standard deviations above the mean (for positive *z* scores) or below the mean (for negative *z* scores) each number in the original batch falls.

PRACTICE

1. Continue to explore the areas of sites near Nanxiong given in Table 3.5 by making box-and-dot plots of Early and Late Bronze Age site areas. How do the levels of the two batches compare?

2. Now compare Early and Late Bronze Age site areas by drawing box-and-dot plots with the levels removed. How do the spreads of the two batches compare?

3. Now compare Early and Late Bronze Age site areas by drawing box-and-dot plots with the levels and spreads removed. How do the two batches compare in terms of symmetry?

4. The largest Early Bronze Age site is 4.2 ha; the largest Late Bronze Age site is 12.8 ha. Which of these sites is more unusual in terms of its batch? Why? Use the median and the midspread of each batch to provide a score for the unusualness of each of these sites. Use the mean and standard deviation of each batch to do the same thing. Do these scores confirm your assessment of which site is more unusual in its batch?

Chapter 5
The Shape or Distribution of a Batch

The *shape* of a batch refers to the way in which the numbers are distributed along the number scale, apart from level and spread. The traditional statistical term for shape is *distribution*. There are two principal aspects to the shape of a batch: *number of peaks* and *symmetry*. We have already discussed batches with multiple peaks and some of the reasons why they must be divided before analysis, so we will proceed directly to the second aspect.

SYMMETRY

Once we have a batch with a single peak, we are in position to use numerical indexes of level and spread. One use to which we can put these numerical indexes is in removing the level and spread so as to evaluate symmetry more carefully. A batch may be symmetrically distributed about its single peak. In a symmetrical batch, about half the numbers fall above the peak, about half the numbers fall below the peak, *and* the numbers above and below the peak stretch away from the peak to similar degrees. That is, the numbers on one side of the peak are no more closely bunched up near the peak than are those on the other side of the peak.

Table 5.1 lists a batch of measurements of volumes of bell-shaped storage pits and illustrates the symmetry of the batch with a stem-and-leaf plot. The stem-and-leaf plot, in fact, shows perfect symmetry. The distribution of numbers above the peak is a perfect mirror image of the distribution of numbers below the peak. The median of this batch is $1.35\,m^3$, and its mean is $1.34\,m^3$. Such close agreement between median and mean is characteristic of batches with symmetrical distributions, and both these numerical indexes of level fall right at the central peak on the stem-and-leaf plot. In short, both behave very well in a symmetrical single-peaked

R.D. Drennan, *Statistics for Archaeologists*, Interdisciplinary Contributions
to Archaeology, DOI 10.1007/978-1-4419-0413-3_5,
© Springer Science+Business Media, LLC 2004, 2009

**Table 5.1. Volumes of Bell-Shaped
Storage Pits at the Buena Vista Site**

Volume (m^3)	Stem-and-leaf plot	
1.23	16	5
1.48	15	15
1.55	14	0568
1.38	13	24589
1.10	12	1349
1.02	11	02
1.29	10	2
1.32		
1.35		
1.65		
1.39		
1.40		
1.12		
1.46		
1.24		
1.34		
1.21		
1.45		
1.51		

batch. They give us exactly the index of the center that matches the pattern that is so clear in the stem-and-leaf plot.

It is unusual to find the perfect symmetry of Table 5.1 in real-world batches of numbers, especially in such small batches as this one. We would be willing to accept a batch this small as symmetrical even if the pattern were considerably less than perfect. Judgments about symmetry are subjective, and we will discuss the process of making them more fully below.

Table 5.2 lists another batch of measurements of volumes of bell-shaped storage pits from a different site. As the stem-and-leaf plot shows, however, this batch is not nearly so symmetrical. Most of the numbers are above the peak, and they tend to stray far above the peak. In contrast, the numbers below the peak are few and lie quite close to the peak. This is an *asymmetrical*, or *skewed*, distribution. Batches can be skewed upward as this one is, or downward if the values tend to stray toward the lower numbers. For discussing symmetry it is especially convenient to draw stem-and-leaf plots with lower numbers at the bottom and higher numbers at the top – like the ones in this book – so that the values in an upwardly skewed shape stray upward on the plot. If your statpack draws them the other way, just remember that when we talk about upward skewness we mean a shape that strays toward the higher numbers, not necessarily toward the top of the stem-and-leaf plot.

Numerical indexes of the center do not behave well at all for a skewed distribution. The median for this batch is $1.29\,m^3$, and the mean is $1.35\,m^3$. These two indexes differ more than did the median and mean for the batch in Table 5.1. More

Table 5.2. Volumes of Bell-Shaped
Storage Pits at the Buenos Aires Site

Volume (m³)	Stem-and-leaf plot	
1.22	20	3
1.64	19	
1.16	18	4
1.07	17	
1.50	16	4
1.84	15	0
1.37	14	03
1.15	13	27
1.29	12	269
1.32	11	1567
2.03	10	47
1.17		
1.04		
1.43		
1.11		
1.40		
1.26		

important, both fall too high on the number scale to accurately reflect the clear single peak at about 1.1 m³. The effect of a skewed distribution on numerical indexes is quite similar to the effect of outliers, as discussed in Chapter 2. Indeed, it is sometimes difficult to tell whether we are looking at a stem-and-leaf plot containing outliers or one showing a skewed distribution. Even the median, highly resistant to the effects of outliers, is affected by a skewed distribution since skewing consists not just of a few aberrant measurements but rather of a pervasive tendency in the shape of the batch.

Since we need a numerical index of the level and spread of a batch in order to begin virtually any statistical analysis, such asymmetrical shapes present us with a serious impediment. Sometimes using the trimmed mean and trimmed standard deviation can help, but it is really the effect of outliers that these indexes eliminate nicely. More fundamental remedies are usually called for before working with a badly asymmetrical shape.

TRANSFORMATIONS

We have already seen that we can perform at least some kinds of arithmetic operations on all the numbers in a batch to produce a new batch that is more amenable to certain kinds of examination. For example, we subtracted the median or the mean from all the numbers in a batch to produce a new batch with the level removed. This had the effect of setting the center to a standard value (zero), while the spread and shape of the batch remained the same. Then we divided all numbers in the zero-level batch by the midspread or the standard deviation to remove the spread. This had the effect of setting the spread to a standard value of one, while the shape of the batch remained the same. *Transformations* are a way of removing the shape of a batch or setting it to a standard shape (single-peaked and symmetrical).

The operations of removing the level and spread are related to each other: first we remove the level, then the spread. We do not remove the spread from a batch without removing the level first. Transformations of shape, however, are independent of removing level and spread. Such transformations are usually performed on batches without removing the level or spread, although they could also be applied after removal of level and spread. Figure 5.1 illustrates the effects that several commonly used transformations have on the shape of a batch. Each batch of numbers is accompanied by a stem-and-leaf plot and by a box-and-dot plot with the level and spread removed. The box-and-dot plots provide the most sensitive indication of symmetry in the original batch and in its various transformations.

x	$\sqrt{x}$	$\log(x)$	$-\dfrac{1}{x}$	$-\dfrac{1}{x^2}$	x^2	x^3	x^4
1.230	1.109	0.207	−0.813	−0.661	1.513	1.861	2.289
1.480	1.217	0.392	−0.676	−0.457	2.190	3.242	4.798
1.550	1.245	0.438	−0.645	−0.416	2.403	3.724	5.772
1.380	1.175	0.322	−0.725	−0.525	1.904	2.628	3.627
1.100	1.049	0.095	−0.909	−0.826	1.210	1.331	1.464
1.020	1.010	0.020	−0.980	−0.961	1.040	1.061	1.082
1.290	1.136	0.255	−0.775	−0.601	1.664	2.147	2.769
1.320	1.149	0.278	−0.758	−0.574	1.742	2.300	3.036
1.350	1.162	0.300	−0.741	−0.549	1.823	2.460	3.322
1.650	1.285	0.501	−0.606	−0.367	2.723	4.492	7.412
1.390	1.179	0.329	−0.719	−0.518	1.932	2.686	3.733
1.400	1.183	0.336	−0.714	−0.510	1.960	2.744	3.842
1.120	1.058	0.113	−0.893	−0.797	1.254	1.405	1.574
1.460	1.208	0.378	−0.685	−0.469	2.132	3.112	4.544
1.240	1.114	0.215	−0.806	−0.650	1.538	1.907	2.364
1.340	1.158	0.293	−0.746	−0.557	1.796	2.406	3.224
1.210	1.100	0.191	−0.826	−0.683	1.464	1.772	2.144
1.450	1.204	0.372	−0.690	−0.476	2.103	3.049	4.421
1.510	1.229	0.412	−0.662	−0.439	2.280	3.443	5.199

Stem-and-leaf plots:

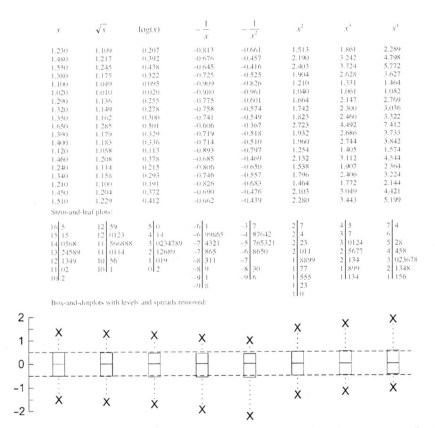

Box-and-dotplots with levels and spreads removed:

Figure 5.1. The effect of transformations on the shape of the batch of measurements from Table 5.1.

Logarithms

The logarithm of a number is the power to which some base must be raised to produce the number. For example, the base-10 logarithm of 1000 is 3 since $10^3 = 1000$. The base-10 logarithm of 100 is 2, since $10^2 = 100$. The base-10 logarithm of 10 is 1, since $10^1 = 10$. We do not usually raise numbers to fractional powers in simple mathematics, but it can be done. Since $10^2 = 100$ and $10^3 = 1000$, $10^{2.14}$ must be greater than 100 and less than 1000. In fact, $10^{2.14} = 137.2$, so the base-10 logarithm of 137.2 is 2.14. One of the vexing chores of introductory statistics always used to be learning to use a table of logarithms. Fortunately, technology has made logarithm tables obsolete, and we can now assume that logarithm transformations will be done with computers or calculators.

The numbers of the third column of Figure 5.1 are actually *natural logarithms*, or base e logarithms. The mathematical constant e has a value of approximately 2.7182818. Its useful characteristics in theoretical mathematics are not of importance to us here, but the logarithms used in many statpacks are base-e logarithms. Thus the numbers in the third column are the powers to which e must be raised to produce the numbers in the original batch. The first number in the original batch, for example, is 1.230. Since $2.7182818^{.207} = 1.230$, the natural logarithm of 1.230 is .207, and .207 appears first in the third column.

Looking first at the original batch of numbers (x in Fig. 5.1), the stem-and-leaf plot shows perfect symmetry (as it did in Table 5.1). The box-and-dot plot confirms this impression.

The transformed batch in the second column of Fig. 5.1 is produced by taking the square root of each of the numbers in the original batch. (See for example, the first number: $\sqrt{1.230} = 1.109$.) This is commonly referred to as the *square root transformation*. The stem-and-leaf plot and the box-and-dot plot for the square root transformation reveal that this new batch has a recognizable tendency to stray downward from its center. (Compare the midspread box or the two extreme adjacent values to those of the original batch in the box-and-dot plots.) The effect of the square root transformation is always to produce a new batch more strongly skewed downward than the original batch, just as we see in this case.

The transformed batch in the third column of Fig. 5.1 is produced by finding the logarithm of each of the numbers in the original batch. As the box-and-dot plots show, this *logarithm transformation*, or log transformation for short, is skewed downward even more strongly than is the square root transformation. Like the square root transformation, it produces a batch with a more downward skewness than the original batch. The effect of the log transformation in this regard is even stronger than the effect of the square root transformation.

The transformed batch in the fourth column of Fig. 5.1 is produced with the *negative reciprocal transformation* $(-1/x)$. The negative reciprocal of the first number in the batch (1.230) is $-1/1.230 = -0.813$. Like the other transformations discussed, it produces a transformed batch with a more pronounced downward skewing than the original batch. Its effect is even stronger than that of the other transformations, as can be seen in the box-and-dot plots at the bottom of Fig. 5.1.

The fifth column of Fig. 5.1 shows an even stronger effect in the same direction. This transformation $(-1/x^2)$ produces downward skewness to an even greater degree. Using the first number again, as an example of the calculation, $-1/1.230^2 = -0.661$. We could continue this progression indefinitely with transformations creating stronger and stronger downward skewness: $-1/x^3$, $-1/x^4$, etc.

Beginning in the sixth column, Fig. 5.1 illustrates transformations that produce the opposite effect. The *square transformation* is simply x^2. (For the first number in the batch, $1.230^2 = 1.513$.) The upward straying effect of the numbers in this small batch after applying the square transformation is barely noticeable.

The *cube transformation* in the seventh column, however, is stronger, and the upward straying of numbers in this transformed batch is easily recognized in the box-and-dot plot. The calculation in this case is simply to raise the original number to the next higher power than in the previous transformation. (For the first number, $1.230^3 = 1.861$.) Even stronger, and in the same positive direction, is the skewing effect of the x^4 transformation in the last column of Fig. 5.1. As with the earlier sequence of transformations producing downward skewing, we could continue this sequence indefinitely to higher and higher powers.

CORRECTING ASYMMETRY

We have just seen how a series of transformations can change the shape of a batch. In this example we started with a batch the shape of which was already symmetrical, and progressively skewed it farther and farther, first in the downward direction, and then in the upward direction. Once we understand them, the effects of the various transformations can be put to good use in changing the shapes of batches that are difficult to work with in the first place because their distributions are not symmetrical. When a transformation producing upward skewing is applied to a batch with downward skewness, the result may be a symmetrical shape.

Precisely the transformations we have just discussed are often used to "correct" asymmetrical shapes. We can use the experience gained in the previous example to list these common transformations and their effects. Table 5.3 summarizes the experience gained from examining the graphs in Fig. 5.1. Or, to put it another way, Fig. 5.1 graphs the practical impact of applying the transformations listed in Table 5.3. Table 5.3 can be used to select an appropriate transformation to apply to an asymmetrical batch like the one in Table 5.2. This batch has a very pronounced tendency to stray upward, so we will need one of the transformations from the lower half of Table 5.3 – transformations that correct upward skewness. The effects of all four of these transformations are illustrated in Fig. 5.2.

Table 5.3. Transformations for Correcting Asymmetry

x^4	Stronger effect	
x^3	Strong effect	Produce upward skewness, that is, correct downward skewness
x^2	Mild effect	
x	No effect	
$\sqrt{x}$	Weak effect	
$\log(x)$	Mild effect	Produce downward skewness, that is, correct upward skewness
$\frac{1}{x}$	Strong effect	
$\frac{1}{x^2}$	Stronger effect	

We easily identified the shape of the batch in the stem-and-leaf plot in Table 5.2 as upwardly skewed. It may be difficult, however, to decide whether a few numbers that stray far from the central bunch represent an upwardly skewed distribution or genuine outliers. The rules of thumb for identifying outliers in box-and-dot plots label the highest two values in the original batch as outliers, as can be seen in the first column of Fig. 5.2. Nevertheless, these rules of thumb, as discussed in Chapter 4, are only arbitrary ways to simplify a complicated relationship between straying numbers and the batch of which they may or may not be a meaningful part. Another approach is to see what effect transformations have on possible outliers.

The weakest transformation for correcting upward straying is the square root transformation, illustrated in the second column of Fig. 5.2. In the transformed batch, the nearer of the two outliers in the untransformed batch no longer qualifies as an outlier; and the box representing the midspread comes closer to being centered on the median. Even disregarding the one outlier still identified, the adjacent values clearly stray farther up than down. The square root transformation produced a less asymmetrical batch, but stronger action is necessary.

The next stronger transformation is the log transformation, illustrated in the third column of Fig. 5.2. In this batch, the median is very close to the center of the midspread. The highest value is still identified as an outlier, but disregarding it, the adjacent values still stray considerably farther upward than downward. The stem-and-leaf plot shows this upward skewness quite clearly. A still stronger transformation is at least worth trying in this instance.

The fourth column in Fig. 5.2 illustrates the effect of the negative reciprocal transformation. The midspread box has now slipped below the middle of the number scale, but the adjacent values still stray farther upward than downward. Most conspicuous, the last remaining outlier no longer qualifies for that status according to the usual rules of thumb. When outliers disappear under the effect of transformations that are also improving the general symmetry of the distribution, it is an indication that they should not be eliminated as outliers but rather treated as straying members of the batch. In such cases, the use of an appropriate transformation is a preferable treatment to correct both asymmetry and apparent outliers. Since the adjacent values continue to stray upward in such a pronounced fashion, it is worth investigating one more transformation with a yet stronger effect.

x	$\sqrt{x}$	$\log(x)$	$-\dfrac{1}{x}$	$-\dfrac{1}{x^2}$
1.220	1.105	0.199	−0.820	−0.672
1.640	1.281	0.495	−0.610	−0.372
1.160	1.077	0.148	−0.862	−0.743
1.070	1.034	0.068	−0.935	−0.873
1.500	1.225	0.405	−0.667	−0.444
1.840	1.356	0.610	−0.543	−0.295
1.370	1.170	0.315	−0.730	−0.533
1.150	1.072	0.140	−0.870	−0.756
1.290	1.136	0.255	−0.775	−0.601
1.320	1.149	0.278	−0.758	−0.574
2.030	1.425	0.708	−0.493	−0.243
1.170	1.082	0.157	−0.855	−0.731
1.040	1.020	0.039	−0.962	−0.925
1.430	1.196	0.358	−0.699	−0.489
1.110	1.054	0.104	−0.901	−0.812
1.400	1.183	0.336	−0.714	−0.510
1.260	1.122	0.231	−0.794	−0.630
Mean: 1.353	1.158	0.285	−0.764	−0.600
Median: 1.290	1.136	0.255	−0.775	−0.601

Stem-and-leaf plots:

```
20 | 3        14 | 3      7 | 1       -4 | 9         -2 | 4
19 |          13 | 6      6 | 1       -5 | 4         -3 | 70
18 | 4        13 |        5 | 0       -6 | 71        -4 | 94
17 |          12 | 8      4 | 1       -7 | 986310    -5 | 731
16 | 4        12 | 03     3 | 246     -8 | 7662      -6 | 730
15 | 0        11 | 578    2 | 0368    -9 | 640       -7 | 643
14 | 30       11 | 124    1 | 0456               -8 | 71
13 | 27       10 | 5788   0 | 47                 -9 | 3
12 | 269      10 | 24
11 | 1567
10 | 47
```

Box-and-dot plots with levels and spreads removed:

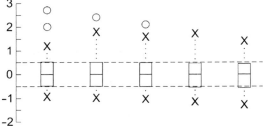

Figure 5.2. Using transformations to correct upward skewness in the batch from Table 5.2.

The fifth column in Fig. 5.2 shows the results of the $-1/x^2$ transformation. The midspread is now less centered on the median than in the previous transformed batch, but the adjacent values have reached a more symmetrical distribution. The stem-and-leaf plot shows about as symmetrical a pattern as it is reasonable to expect in a real-world batch of numbers this small. The decision between the last two transformations is difficult. Both have succeeded in eliminating outliers. The adjacent values look more symmetrical in the $-1/x^2$ transformation, while the midspread looks more symmetrical in the $-1/x$ transformation. We might reasonably use the more symmetrical appearance of the stem-and-leaf plot for the $-1/x^2$ transformation to break the tie and opt for the transformed batch in the last column as the most symmetrical, but either of these batches is symmetrical enough to analyze. That is to say, either batch is symmetrical enough that the mean and standard deviation would be accurate and useful indexes of center and spread.

Transformations often seem an arcane statistical ritual performed more for superstitious reasons than anything else. Their purpose, however, is simply to provide a batch of numbers whose shape makes it possible for the mean and standard

Figure 5.3. Transformation rulers: a "normal" ruler (above); a ruler that would give length measurements with a square transformation (center); the same square transformation ruler with tick marks every five units instead of every 1/5 of the length (bottom).

deviation to be useful indexes of the center and the spread. The mean and standard deviation are fundamental to many of the techniques discussed farther on in this book, and if the mean and standard deviation are not telling us the truth about the center and spread of the batch, then those other techniques will not work well. Transformation can be thought of as measuring with special rulers. Fig. 5.3 shows three rulers. At the top is a "normal" ruler that we might use to measure the length of some object. In the middle is a "square transformation ruler." If we measured the same object with this ruler, the result would be the square of the "normal" measurement. The bottom ruler is a square transformation ruler just like the middle ruler, except that the tick marks, instead of being evenly spaced along the ruler, are placed every five units. This shows the way the units of measurement are distributed differently along the square transformation ruler than they are along the "normal" ruler, and may provide a better common-sense feel for just how it is that the square transformation shifts numbers along the scale to change the shape of a batch. The same is true of all the other transformations we have discussed. Using them amounts to nothing more than measuring with a peculiar ruler. Although it certainly seems strange at first thought, there is no reason that we couldn't use rulers like the square transformation rulers in Fig. 5.3. We would just have to measure everything we wanted to compare with the same peculiar ruler. That is exactly what we are doing when we transform a batch of numbers – we are measuring with a peculiar ruler, and we must use the same peculiar ruler (or transformation) on all the batches we want to be able to compare.

THE NORMAL DISTRIBUTION

The single-peaked symmetrical shape we have been pursuing with transformations in this chapter has the essential characteristics of the (in)famous *normal distribution*. Actually, the normal distribution implies some other more specific characteristics, but in practical terms, a batch that is single peaked and symmetrical can be taken as close enough to a normal distribution to apply statistical techniques suitable for batches that are normally distributed.

The requirement of normality in the distributions of batches that are to be analyzed in certain ways is no deep mathematical mystery. It is not a question of abiding by some secret and sacred principle understandable only to the high priests of statistics. Understanding the importance of the normal distribution begins with such simple and intuitively intelligible notions as the ways in which numerical indexes of level and spread work on asymmetrical batches. We have seen that these numerical indexes simply do not produce sensible results when applied to a batch that is not single peaked and symmetrical. This is the starting point for understanding why some statistical techniques must only be applied to normal distributions. Many of them begin by characterizing the batch to be analyzed with the mean and standard deviation. If these numerical indexes do not provide accurate and meaningful measures of the center and spread of the batch, then no technique that takes them as a starting point can be expected to produce accurate and meaningful results.

To summarize, then, if we wish to study a batch of numbers that is not single peaked and symmetrical, we must often take special action. This consists, first, of splitting a batch with multiple peaks into multiple separate batches, each with a single peak. Second, we can use transformations to make the shape of a single-peaked batch more symmetrical and/or deal with any outliers we observe. These initial data preparation steps are important to the success of many statistical techniques and must not be overlooked. Analysis of a batch of numbers should *always* begin by exploring the batch with a stem-and-leaf plot, and taking whatever corrective action may be indicated to deal with multiple peaks, asymmetrical shapes, and outliers.

Picking the best transformation to correct asymmetry is a question of subjective judgment. It requires a bit of practice to look at the distributions produced by different transformations and decide which is most symmetrical. These judgments are especially difficult in small batches of numbers where displacing a single number in the stem-and-leaf plot can have a strong effect on the apparent symmetry. It is a good idea not to be too strongly swayed by appearances that could be changed if only one or two numbers in the batch were slightly different. It is better, instead, to concentrate more heavily on major trends that would only be altered if many numbers were changed.

Picking the best transformation is also a process of trial and error. Although Table 5.3 might help you guess which transformation to try, it is almost always necessary to try several transformations and look at the results (by examining stem-and-leaf and box-and-dot plots of the transformed batches) in order to decide which produces the most symmetrical shape. Compromises are often required, especially when transformations are being applied to two or more batches that eventually are to be compared. The same transformation may not produce the most symmetrical shape for each of the batches involved, but the same transformation must be applied to all the batches if they are to be compared after transformation.

An alternative to correcting asymmetry through transformations is to use statistical techniques based not on the mean and standard deviation but on other indexes – like the trimmed mean and trimmed standard deviation – that are more resistant to the effects of outliers and asymmetry. Such approaches will be discussed where relevant in the following chapters. Generally, if the presence of outliers presents a problem for the use of means and standard deviations, use of the trimmed mean and standard deviation is a good solution. If pervasive asymmetry in the distribution is the problem, applying an appropriate transformation is more effective.

PRACTICE

1. Look carefully at the shapes of the batches of areas of Early and Late Bronze Age sites near Nanxiong from Table 3.5. In the practice questions from Chapters 3 and 4, you have already made stem-and-leaf and box-and-dot plots of these batches. Does either batch have a skewed shape? If so, is it skewed upward or downward?
2. If either Early or Late Bronze Age site areas are skewed, use your statpack to experiment with transformations to correct the asymmetry. Which transformation would you choose for each batch individually? Why? Which transformation would you use for both if you intended to compare the transformed batches? Why?

Chapter 6
Categories

The batches of numbers that we have discussed in Chapters 1–5 have all consisted of a set of measurements of some kind. There is a fundamentally different kind of batch that we must discuss before continuing with the second section of this book. This other kind of batch results from observations of characteristics that are not measured exactly, but instead are grouped into different *categories*. Archaeologists are quite accustomed to the notion of categorizing things. We usually discuss it under the heading of typology, and the definition of typologies (or sets of categories) for artifacts is widely recognized as a fundamental initial step in description and analysis. Much has been written about the "correct" way to define pottery types in particular. Our concern here is not how to define categories, but rather what to do with the result once we have defined them and counted up how many things there are in each one. When we classify the ceramics from a site as Fidencio Coarse, Atoyac Yellow-White, and Socorro Fine Gray, we are dealing in categories. When we count the number of flakes, blades, bifaces, or debitage from a site, we are also dealing in categories. When we add up the number of cave sites and open sites in a region, we are once again dealing in categories. When we divide the sites in a region into large sites, medium-sized sites, and small sites, we are still once more dealing in categories. Data recorded in terms of such categories comprise batches just as do data recorded as true measurements (for example, in centimeters, grams, hectares, etc.).

Table 6.1 provides an example of such categorical information for a set of 140 pottery sherds. One observation made about each sherd is the site where it was recovered. There are three categories here: the Oak Grove site, the Maple Knoll site, and the Cypress Swamp site. A second observation concerns incised decoration, with two categories: each sherd is either incised or unincised. This may seem an unusual way to present such information, and indeed it is. The presentation is

R.D. Drennan, *Statistics for Archaeologists*, Interdisciplinary Contributions
to Archaeology, DOI 10.1007/978-1-4419-0413-3_6,
© Springer Science+Business Media, LLC 2004, 2009

Kinds of Data

Some statistics books begin with a chapter about different kinds of data as if the recognition of a standard set of several fundamentally different kinds of data were the rock upon which all statistical analysis was built. The "fundamentally" different kinds, however, are defined in different ways by different authors, and many books don't make much of such distinctions at all. There are almost as many sets of terms as there are authors, and some of the same terms are used in contradictory ways by different authors. The point is that there are a number of characteristics of batches of numbers that vary. You can analyze batches with certain characteristics in one way, but batches that lack those characteristics can be analyzed in a different way. The most important distinction made in this book is between what are here called *measurements* and *categories*.

Measurements are things like lengths, widths, areas, weights, and so on—quantities we measure along a scale of appropriate units. True measurements come in fractional values as well as whole-number values (difficult numbers to work with like $3\text{-}^{13}/_{16}$ inches, or much easier numbers to work with like 9.68 cm), and there is, in principle at least, an infinite number of potential values along the scale. Measurements, as the term is used in this book, can also include numbers of things, like number of inhabitants in different regions, number of artifacts in different sites, and so on. Measurements can also be derived from other measurements arithmetically. Densities of artifacts in different excavation units, for example, are measurements derived by dividing the number of artifacts encountered in each excavation unit by the volume of deposit excavated to arrive at the number of artifacts per cubic meter for each excavation unit. In exploratory data analysis, measurements are sometimes called *amounts* (measurements along a scale) and *counts* (counted numbers of things) and *balances* (which can have positive and negative values). Measurements are made along what are sometimes called *ratio* scales or *interval* scales. A ratio scale has a meaningful zero point, as, for example, a length or weight. An interval scale has an arbitrary zero point, which prevents some kinds of manipulations. The usual example of a scale with an arbitrary zero point is temperature. The fact that the zero point is arbitrary means that you really can't say that 60° is twice as hot as 30° (on either Fahrenheit or Celsius scales, although you can do such things with the temperature scale measured from absolute zero).

Categories are essentially groups of things, and we count the number of things in each group. We ordinarily work with sets of categories that are *mutually exclusive and exhaustive*. That is, each thing in the set of things we are studying must fit into one category and only one category. Pottery types are a common kind of category in archeology, and we recognize that pottery types need to be defined so that each sherd can be placed in one and only one type. Colors represent another set of categories. We may sort things out as red, blue, or green. If we find bluish-green things, we may need to add a fourth category to the set. Categories are sometimes called *nominal data*. If the categories can logically be arranged in a specific order, then they form *ordinal data* or *ranks*. Pottery types do not have this property. Categories like *large, medium, small,* and *tiny* do; we recognize that to say small, large, tiny, medium is to put these categories out of order. If there are very many categories in the set, ranks begin to act like true measurements in some ways.

The most important distinction between kinds of data for the organization of thoughts in this book is between what we will call measurements and categories, but we will also consider some special treatments that can be applied to ranks—treatments that relate strongly to things we can do with true measurements.

---

cumbersome and tells us virtually nothing about patterns. In an instance like this, we would much more likely present the information as a *tabulation*, and that is what we will do shortly. It is often convenient to manage categorical information in the manner in which it is presented in Table 6.1, however, especially with computers. Thus it is important to recognize Table 6.1 as one means of organizing the same information we will see in more familiar form in the following tables. Table 6.1 is the most complete and detailed way of recording this information and the most similar to the way in which batches of measurements were initially presented in previous chapters.

Table 6.2 presents the information about where the sherds were recovered in a more compact, familiar, and meaningful way. This simple tabulation of *frequencies* (or *counts*) and *proportions* (or *percentages*) immediately tells us something about how much pottery came from where – something that was not at all apparent in Table 6.1. More pottery came from Oak Grove than any other site and less from Maple Knoll. Table 6.3 performs the same task for the information about decoration for the same 140 sherds. Most of the sherds are unincised, but the difference in proportions between incised and unincised is not extreme.

In effect, Table 6.1 contains sets of related batches, like the related batches that have been discussed in previous chapters. In this case, we could divide the sherds into three related batches, as in Table 6.2: sherds from the Oak Grove site, sherds from the Maple Knoll site, and sherds from the Cypress Swamp site. Or we could divide them in a different way into two related batches, as in Table 6.3: incised sherds and unincised sherds. Each set of categories is simply one way of dividing the whole set of sherds into different batches. We might well want to compare the first three batches (the sherds from each of the three sites) in regard to the other set of categories (incised decoration). Table 6.4 extends the tabulations of Tables 6.2 and 6.3 to accomplish this comparative goal, by simultaneously dividing the sherds

Table 6.1. Information about 140 Pottery Sherds

Oak Grove	Unincised	Maple Knoll	Unincised	Cypress Swamp	Incised
Maple Knoll	Incised	Oak Grove	Incised	Cypress Swamp	Unincised
Cypress Swamp	Unincised	Oak Grove	Unincised	Cypress Swamp	Unincised
Cypress Swamp	Incised	Oak Grove	Unincised	Oak Grove	Incised
Cypress Swamp	Incised	Maple Knoll	Incised	Oak Grove	Unincised
Cypress Swamp	Unincised	Cypress Swamp	Unincised	Maple Knoll	Incised
Cypress Swamp	Incised	Cypress Swamp	Unincised	Maple Knoll	Unincised
Oak Grove	Incised	Oak Grove	Incised	Oak Grove	Incised
Oak Grove	Unincised	Oak Grove	Unincised	Oak Grove	Unincised
Maple Knoll	Unincised	Maple Knoll	Unincised	Maple Knoll	Incised
Oak Grove	Incised	Cypress Swamp	Incised	Cypress Swamp	Incised
Oak Grove	Unincised	Cypress Swamp	Incised	Cypress Swamp	Unincised
Maple Knoll	Incised	Oak Grove	Incised	Oak Grove	Incised
Maple Knoll	Unincised	Oak Grove	Unincised	Oak Grove	Unincised
Cypress Swamp	Unincised	Maple Knoll	Unincised	Oak Grove	Unincised
Cypress Swamp	Incised	Cypress Swamp	Unincised	Maple Knoll	Incised
Oak Grove	Unincised	Cypress Swamp	Incised	Cypress Swamp	Unincised
Maple Knoll	Incised	Oak Grove	Incised	Cypress Swamp	Incised
Maple Knoll	Unincised	Oak Grove	Unincised	Oak Grove	Incised
Cypress Swamp	Incised	Maple Knoll	Unincised	Oak Grove	Unincised
Oak Grove	Incised	Cypress Swamp	Unincised	Maple Knoll	Unincised
Oak Grove	Unincised	Cypress Swamp	Incised	Maple Knoll	Incised
Maple Knoll	Unincised	Oak Grove	Unincised	Oak Grove	Incised
Cypress Swamp	Unincised	Maple Knoll	Incised	Oak Grove	Unincised
Oak Grove	Incised	Cypress Swamp	Unincised	Cypress Swamp	Incised
Oak Grove	Unincised	Oak Grove	Unincised	Cypress Swamp	Unincised
Maple Knoll	Incised	Maple Knoll	Incised	Cypress Swamp	Unincised
Maple Knoll	Unincised	Cypress Swamp	Unincised	Oak Grove	Incised
Oak Grove	Incised	Cypress Swamp	Incised	Oak Grove	Unincised
Oak Grove	Unincised	Oak Grove	Incised	Maple Knoll	Unincised
Maple Knoll	Incised	Oak Grove	Unincised	Cypress Swamp	Incised
Cypress Swamp	Incised	Oak Grove	Unincised	Oak Grove	Incised
Cypress Swamp	Unincised	Maple Knoll	Incised	Oak Grove	Unincised
Oak Grove	Incised	Cypress Swamp	Unincised	Oak Grove	Unincised
Oak Grove	Unincised	Oak Grove	Incised	Maple Knoll	Incised
Maple Knoll	Incised	Oak Grove	Unincised	Cypress Swamp	Unincised
Maple Knoll	Unincised	Oak Grove	Unincised	Oak Grove	Incised
Cypress Swamp	Unincised	Maple Knoll	Incised	Oak Grove	Unincised
Maple Knoll	Incised	Cypress Swamp	Unincised	Maple Knoll	Unincised
Cypress Swamp	Unincised	Oak Grove	Incised	Cypress Swamp	Unincised
Oak Grove	Incised	Oak Grove	Unincised	Cypress Swamp	Incised
Oak Grove	Unincised	Maple Knoll	Incised	Cypress Swamp	Unincised
Maple Knoll	Unincised	Cypress Swamp	Incised	Oak Grove	Incised
Cypress Swamp	Unincised	Cypress Swamp	Unincised	Oak Grove	Unincised
Maple Knoll	Incised	Oak Grove	Incised	Maple Knoll	Incised
Oak Grove	Incised	Oak Grove	Unincised	Maple Knoll	Unincised
Oak Grove	Unincised	Maple Knoll	Incised		

Calculating Percentages and Rounding Error

We have noted *rounding error* before, but Tables 6.2 and 6.3 provide an opportunity to clear up this little mystery completely. We know that 140 sherds are 100%, and the percentages in Table 6.3 add up to 100.0, but the percentages of the three categories in Table 6.2 add up to only 99.9%. In both tables the percentages have been rounded off to one digit following the decimal point. For Table 6.3, the full calculations of the percentages are 64 / 140 = .4571428571428571428... and 76 / 140 = .5428571428571428571...

Both of these numbers will continue to repeat the same sequence of digits (...142857...) forever. The division will never come out even, no matter how far out it is carried. To change .4571428571428571428... and .5428571428571428571... from ordinary decimal fractions into percentages, of course, we multiply them by 100: 45.71428571428571428...% and 54.28571428571428571...%. (And while we're on the subject, it is worth emphasizing that 0.45 and 45.0% are the same number. There is a big difference between 0.45 and 0.45%—0.45 means .45 out of 1.00 or 45 out of 100 or 4,500 out of 10,000 [that is, almost half], but 0.45% means 0.45 out of 100 or 45 out of 10,000 [or far less than half]. It is essential to be careful with a decimal point and the % symbol.) Clearly the percentages we have here must be rounded off. If we want one digit after the decimal place in the percentage, 45.71428571428571428...% rounds *down* to 45.7% and we lose the extra .01428571428571428...%, and 54.28571428571428571...% rounds *up* to 54.3%. In rounding this number up we have actually added .01428571428571428... to it, which is exactly the amount that we lost in rounding the other percentage down. Since one percentage has been raised by the same amount that the other has been lowered by, the amounts cancel each other out when we add the percentages together, and the total is 100.0%.

In Table 6.2, however, all three percentages turn out to round down:

- 59/140 = .42142857142857... which rounds down to .421 (or 42.1%), losing .042857142857...%;
- 37/140 = .26428571428571... which rounds down to .264 (or 26.4%), losing .028571428571...%; and
- 44/140 = .31428571428571... which rounds down to .314 or 31.4%, losing .028571428571...%.

If we add up what we have lost in rounding all three percentages down, we get almost exactly the 0.1% that is missing from the 99.9% total of the three rounded off percentages:

$$(.042857142857...\% + .028571428571...\% + .028571428571...\% = .099999999999...\%)$$

Precisely the same thing can happen in the other direction if more is gained by rounding some percentages up than is lost by rounding others down. Thus the total of a set of percentages can be slightly more than 100%. Sometimes, doing percentage calculations with more decimal digits of precision will remove rounding error. In a case like this example, however, where the quotient of the division repeats infinitely, it doesn't matter how far out we carry the calculations. They will never come out *exactly* even. Sooner or later we have to round off, and accept a little rounding error.

Table 6.2. Sherds from Three Sites

	Oak Grove	Maple Knoll	Cypress Swamp	Total
Frequency	59	37	44	140
Proportion	42.1%	26.4%	31.4%	99.9%

Table 6.3. Pottery Decoration

	Incised	Unincised	Total
Frequency	64	76	140
Proportion	45.7%	54.3%	100.0%

Table 6.4. Incised and Unincised Sherds from Three Sites

a. Frequencies

	Oak Grove	Maple Knoll	Cypress Swamp	Total
Incised	25	21	18	64
Unincised	34	16	26	76
Total	59	37	44	140

b. Column Proportions

	Oak Grove	Maple Knoll	Cypress Swamp	Average
Incised	42.4%	56.8%	40.9%	45.7%
Unincised	57.6%	43.2%	59.1%	54.3%
Total	100.0%	100.0%	100.0%	100.0%

c. Row Proportions (not useful in this instance)

	Oak Grove	Maple Knoll	Cypress Swamp	Total
Incised	39.1%	32.8%	28.1%	100.0%
Unincised	44.7%	21.1%	34.2%	100.0%
Average	42.1%	26.4%	31.4%	99.9%

by site and by incised decoration. Such a tabulation is sometimes called a *cross tabulation* or *two-way table*, because it divides the entire set of sherds into categories in two different ways simultaneously. In this kind of table there are also two different ways to use percentages in comparing these batches.

COLUMN AND ROW PROPORTIONS

Following the frequencies in the two-way table are *column proportions* (Table 6.4b). Each of the column proportions is a proportion of the total number of sherds at the corresponding site, so the proportions for each site add up to 100% (within rounding error). These proportions are similar to those in Table 6.3, but now they are calculated separately for each of the three sites. The average column proportions at the extreme right of Table 6.4b are not simply the averages of the individual site proportions. They are actually the same as the proportions in Table 6.3. That is, they are the proportions for the complete set of sherds considering all sites together.

Column proportions are useful for comparing columns to each other – in this instance comparing the three sites to each other with regard to their relative proportions of incised and unincised sherds. The assemblage of sherds from the Maple Knoll site stands out in regard to incised pottery, since 56.8% of the sherds from the Maple Knoll site are incised. At the Oak Grove and Cypress Swamp sites, incised pottery is considerably less abundant, amounting to 42.4% and 40.9% of the sherds at the two sites, respectively.

Table 6.4c provides *row proportions* for the table. These are proportions, not of the total number of sherds from each site, but rather of the total number of sherds in each decoration category (incised and unincised). This is not the way we would want to calculate proportions in this instance. The highest proportion for incised sherds, for example, in Table 6.4c is for the Oak Grove site, but this is at best uninteresting, and at worst misleading. The highest proportion for unincised sherds is also at the Oak Grove site. These high proportions simply reflect the fact that we made a larger collection at the Oak Grove site than at any other (59 sherds, as opposed to 37 and 44 at the other sites, as seen in Table 6.4a). As a consequence we came home with more incised sherds *and* more unincised sherds than from either of the other sites. The prevalence of incised decoration in the assemblage at the Maple Knoll site (reflected meaningfully in the column proportions) is completely obscured in the row proportions because of the probably entirely meaningless circumstance that the collection from Oak Grove consists of more sherds.

We could, of course, have set the two-way table up the other way in the first place, with rows corresponding to sites and columns to decoration categories instead of columns corresponding to sites and rows to decoration categories. If we had done that, then we would have wanted row proportions, not column proportions. Whether to choose row proportions or column proportions for a particular table seems intuitively obvious to many people, but not to everyone. Occasionally there is a situation in which either row proportions or column proportions might be meaningful, depending on which point needs to be made. In the prototypical archaeological situation of calculating proportions of different artifact categories so as to compare the assemblages from different regions, sites, features, strata, and the like, it is always proportions that add up to 100% for each assemblage that we want – *not* proportions that add up to 100% for each artifact category.

PROPORTIONS AND DENSITIES

Proportions are used in a wide variety of contexts, but one context in particular arises over and over again in archaeology: comparing the proportions of different categories in artifact or ecofact assemblages from different contexts or locations. Such comparisons can obviously not be based on the frequencies or counts of artifact categories directly, since we are likely to have many more artifacts from some places than others. A very large number of, say, deer bones from one stratum does not necessarily mean a faunal assemblage especially rich in deer bone; it might mean only a stratum which yielded an especially large amount of bone. Archaeologists have sometimes said that such assemblages must be "standardized" in order to be compared. By this they mean that the varying quantities of things from the different units we want to compare must somehow be equalized. It is better not to call this "standardization" because in statistics this word is already used to mean something else (as discussed in Chapter 4). But it's true, the effect of large numbers of things from some places and small numbers of things from other places must somehow be set aside in order to compare them.

Especially when the assemblages being compared come from different excavation units (strata, features, and the like), archaeologists have often calculated the *densities* of different categories of things by dividing the number recovered by the volume of excavated deposits from which they were recovered. Total densities of artifacts or ecofacts can sometimes be useful, but they do not usually provide a very good basis for comparing the composition of different assemblages to each other. Table 6.5 provides an example to illustrate this point. It details results from five different excavation units in different locations within an archaeological site. As can be seen in the second column, Units 1, 2, and 5 represent relatively small amounts of excavated deposit, quite likely because only a small test excavation was carried out in these locations. Unit 3 was only slightly larger. Unit 4, however,

Table 6.5. Proportions and Densities

Excavation Unit	Volume Excavated	Total Sherd Number	Total Sherd Density	Decorated Sherds			
				Number	% Total Decorated	Density	% Total Assemblage
1	2.3 m³	213	93/m³	18	17%	**7.8/m³**	9%
2	1.7 m³	193	114/m³	16	15%	**9.4/m³**	8%
3	5.1 m³	39	8/m³	20	19%	3.9/m³	**51%**
4	21.2 m³	1483	70/m³	**37**	**36%**	1.7/m³	3%
5	1.6 m³	433	271/m³	13	13%	**8.1/m³**	3%

represents substantially more excavation, and, not surprisingly, far more sherds were recovered from Unit 4 than from any other (the third column).

As the fourth column in Table 6.5 shows, the densities of artifacts in the deposits excavated also varied substantially, from extremely dense for Unit 5 to extremely sparse for Unit 3. As a consequence, although Unit 3 represents the second largest volume of excavated deposit, it yielded the smallest number of sherds. The fact that sherds of whatever variety were very dense in Unit 5 may well reflect very intensive utilization of that location by the site's ancient inhabitants.

If our attention turns to comparison of the composition of the ceramic assemblages of these five locations, however, the fact that some excavation units were very large, and some very small, gets in the way. So does the fact that sherd densities were very high in some units and very low in others. No one would suggest that Unit 4 stands out for the prevalence of decorated sherds simply because more decorated sherds were recovered there. It seems self-evident that the large number of decorated sherds recovered there (37) are attributable to the large size of the excavation unit. The proportions in the sixth column tell us nothing more. That 36% of the decorated sherds recovered came from Unit 4 reveals nothing more than that Unit 4 was a large excavation. These proportions are not meaningful, as discussed in the previous section.

In very similar fashion, the densities of decorated sherds in the seventh column provide a very misleading view of where decorated ceramics were most prevalent. Units 1, 2, and 5 all have quite high densities of decorated sherds, but this is telling us nothing more than that these units have high densities of all kinds of sherds (look at the fourth column).

It is the last column that says what needs to be said about where decorated ceramics were most prevalent. These are proportions that add up to 100% for the sherds in each excavation unit's assemblage. The salient fact is that over half the sherds recovered from Unit 3 were decorated, whereas fewer than 10% were in all the other units. Clearly, ceramics were much more decorated at this location than at the other four locations. The effects of differing excavated volumes and of differing sherd densities are set aside effectively for comparison by these proportions. Calculating densities of decorated sherds does not accomplish this aim. For comparing assemblages with regard to their constituent categories of things, then, we want to look at the categories as proportions of the assemblages they come from.

BAR GRAPHS

The bar graph, a relative of the histogram, provides a familiar way to show proportions graphically. Both bar graphs in Fig. 6.1 illustrate the column proportions we have just discussed. They differ only in the way the bars are grouped. The bar graph at the left groups together the three bars representing the proportions of incised sherds at each of the three sites, and then does the same for the unincised sherds. The bar graph at the right groups together the two bars representing the proportions

Figure 6.1. Bar graph of proportions of incised and unincised sherds at the Oak Grove, Maple Knoll, and Cypress Swamp sites.

of incised and unincised sherds at each site. Most often, when we draw bar graphs representing the proportions of artifacts in different categories at archaeological sites, it makes sense to group all the bars for one site assemblage together, as in the bar graph at the right in Fig. 6.1. Each group of bars then represents visually the makeup of a single assemblage, reflecting the basis on which the proportions were calculated (percentages of the different categories within each assemblage). This is usually the most effective way to present the differences between assemblages of artifacts, which is what we are most likely interested in. This configuration of bars calls attention to the fact that the assemblages from Oak Grove and Cypress Swamp are quite similar, but the assemblage of Maple Knoll differs because of its high proportion of incised sherds.

Many computer programs make it easy to produce bar graphs that are visually much more arresting than those in Fig. 6.1. We can consider the somewhat more complex example of proportions of eight ceramic types (A–H) at four sites (Oak Grove, Maple Knoll, Cypress Swamp, and Cedar Ridge). These are illustrated in Fig. 6.2 in the same way that the proportions of incised and unincised sherds from

Figure 6.2. Bar graph of proportions of eight ceramic types in the assemblages from the Oak Grove, Maple Knoll, Cypress Swamp, and Cedar Ridge sites.

Figure 6.3. Pseudo three-dimensional bar graphs representing the same proportions as Fig. 6.2, but less clearly.

some of the same sites were illustrated in Fig. 6.1. The approximate proportion of each type in each assemblage can be read relatively easily from Fig. 6.1. Beyond this, a broad similarity of assemblage composition between the Oak Grove and Cypress Swamp sites is apparent, while the Maple Knoll and Cedar Ridge ceramic assemblages are rather different from this pattern, and from each other. The pseudo three-dimensional effect of the bar graphs in Fig. 6.3 introduces visual clutter that makes them more difficult to read than the simpler flat bar graphs of Fig. 6.2. The stacked bars of Fig. 6.4 produce a cacophony of visual noise and convey very little information. Carrying the bar graphs fully into three dimensions as in Fig. 6.5 almost completely obscures anything they might have illustrated. Although pie charts are often used to illustrate the proportions of the categories in a whole, Fig. 6.6 shows how much more difficult they make it to recognize the patterns that are fairly obvious in Fig. 6.2. Less is more.

CATEGORIES AND SUB-BATCHES

Categories enable us to break a batch down into sub-batches which can then be compared to each other. The comparison may be of another set of categories, as in the example in this chapter, or it may be of a measurement. If, for example, we measured the approximate diameter of the vessel represented by each of the sherds

Figure 6.4 Stacked bar
graph representing the same
proportions as Fig. 6.2, but
much less clearly.

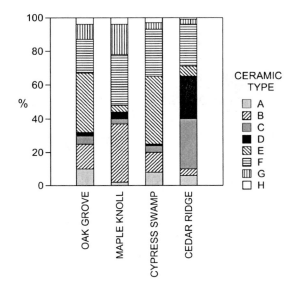

Figure 6.5 Bar graph in
three dimensions making the
patterns visible in Fig. 6.2
impossible to see.

OAK GROVE   MAPLE KNOLL   CYPRESS SWAMP   CEDAR RIDGE

Figure 6.6. Pie charts representing the same proportions as Fig. 6.2, but very poorly.

from Table 6.1, we could then break the batch of sherds into sub-batches according to site and compare vessel diameters for the three sites. The tools needed for that comparison are precisely those already discussed in Chapters 1–4. We could, for example, draw three box-and-dot plots (one for the sub-batch representing each site) all at the same scale to compare rim diameters of vessels at the three sites.

PRACTICE

1. Beginning to assess settlement distribution in the area around Al-Amadiyah, you select 400 random points on the map of your study area and visit each one in the field. You classify each of the 400 points according to its setting (alluvial valley floor, rocky piedmont, or steeper mountain slopes), and you observe whether there is any evidence of prehistoric occupation there. Your results are as follows:

   - 41 points are in the alluvial valley floor. Of these, 14 show evidence of prehistoric occupation, and 27 do not.
   - 216 points are in the stony piedmont. Of these, 64 show evidence of prehistoric occupation, and 152 do not.
   - 143 points are on the steeper mountain slopes. Of these, 20 show evidence of prehistoric occupation, and 123 do not.

   Use proportions and a bar chart to compare the three environmental settings in regard to the density of prehistoric occupation your preliminary field work found in each. Would you say that some zones were more filled than others by prehistoric inhabitants? If so, which one(s)?
2. You continue your study at Al-Amadiyah by revisiting each of the 98 locations that did show evidence of prehistoric occupation, and you measure the areal extent of the surface scatters of artifacts that indicate the archaeological sites. Your results are presented in Table 6.6. Use the environmental settings to separate the 98 locations into three separate batches, and use box-and-dot plots to compare the three batches in regard to site area. Do site sizes appear to differ from one setting to another? Just how?

**Table 6.6. Areas of Sites in Three Environmental Settings in
the Study Area at Al-Amadiyah**

Site area (ha)	Setting	Site area (ha)	Setting
2.8	Piedmont	2.5	Piedmont
7.2	Piedmont	2.0	Piedmont
3.9	Piedmont	8.8	Alluvium
1.3	Slopes	20.3	Alluvium
2.3	Piedmont	5.5	Piedmont
6.7	Piedmont	3.5	Piedmont
3.0	Piedmont	8.3	Piedmont
2.3	Piedmont	6.4	Piedmont
4.2	Piedmont	4.1	Piedmont
0.4	Slopes	0.8	Slopes
3.5	Piedmont	0.7	Slopes
2.7	Piedmont	7.7	Piedmont
19.0	Alluvium	5.8	Piedmont
6.0	Piedmont	2.9	Piedmont
4.5	Piedmont	4.8	Piedmont
2.9	Slopes	4.9	Piedmont
5.3	Piedmont	1.0	Slopes
4.0	Piedmont	2.3	Piedmont
3.3	Piedmont	1.5	Slopes
0.8	Slopes	9.3	Alluvium
7.7	Alluvium	2.9	Piedmont
2.6	Piedmont	1.1	Piedmont
1.5	Piedmont	0.8	Slopes
4.2	Piedmont	1.9	Piedmont
15.8	Alluvium	6.9	Piedmont
4.7	Piedmont	0.9	Slopes
2.1	Piedmont	9.8	Piedmont
1.4	Slopes	6.2	Alluvium
1.1	Slopes	7.4	Piedmont
8.1	Piedmont	3.6	Piedmont
4.2	Piedmont	3.2	Piedmont
1.2	Slopes	7.3	Piedmont
6.7	Alluvium	0.5	Slopes
8.5	Piedmont	2.1	Piedmont
3.0	Piedmont	0.7	Slopes
5.3	Piedmont	3.1	Piedmont
10.5	Alluvium	4.5	Piedmont
2.3	Piedmont	2.0	Slopes
4.1	Piedmont	17.7	Alluvium
10.2	Alluvium	5.7	Piedmont
9.3	Alluvium	5.2	Piedmont
3.4	Piedmont	2.2	Piedmont
7.7	Alluvium	0.5	Slopes
8.8	Piedmont	2.4	Piedmont
7.9	Piedmont	2.0	Piedmont
4.9	Alluvium	2.5	Piedmont
3.7	Piedmont	5.3	Piedmont
1.3	Slopes	0.3	Slopes
3.2	Piedmont	1.0	Slopes

Part II
Sampling

Chapter 7
Samples and Populations

The notion of sampling is at the very heart of the statistical principles discussed in this book, so it is worth pausing here at the beginning of Part II to discuss clearly what sampling is and to consider some of the issues that the practice of sampling raises in archaeology. Archaeologists have, in fact, been practicing sampling in one way or another ever since there were archaeologists, but widespread recognition of this fact has only come about in the past 20 years or so. In 1970 the entire literature on sampling in archaeology consisted of a very small handful of chapters and articles. Today there are hundreds and hundreds of articles, chapters, and whole books, including many that attempt to explain the basics of statistical sampling to archaeologists who do not understand sampling principles.

Unfortunately, many of these articles seem to have been written by archaeologists who do not themselves understand the most basic principles of sampling. The result has been a great deal of confusion. It is possible to find in print (in otherwise respectable journals and books) the most remarkable range of contradictory advice on sampling in archaeology, all supposedly based on clear statistical principles. At one extreme is the advice that taking a 5% sample is a good rule of thumb for general practice. At the other extreme is the advice that sampling is of no utility in archaeology at all because it is impossible to get any relevant information from a sample or because the materials archaeologists work with are always incomplete collections anyway and one cannot sample from a sample. (For reasons that I hope will be clear well before the end of this book, both these pieces of advice are wrong.) It turns out that good sampling practice requires not the memorization of a series of arcane

R.D. Drennan, *Statistics for Archaeologists*, Interdisciplinary Contributions to Archaeology, DOI 10.1007/978-1-4419-0413-3_7,
© Springer Science+Business Media, LLC 2004, 2009

rules and procedures but rather the understanding of a few simple principles and the thoughtful application of considerable quantities of common sense.

There is another way, too, in which a pause for careful consideration of sampling principles can be useful to archaeologists. A lament about the regrettably small size or the questionable representativeness of the sample is a common conclusion to archaeological reports. Statisticians have put a great deal of energy into thinking about how we can work with samples. Some of the specific tools they have developed could be used to considerably more advantage in archaeology than they often have been in the past, and this book attempts to introduce several of them. More fundamentally, though, at least some of the logic of working with samples on which statistical techniques are based is equally relevant to other (nonstatistical) ways of making conclusions from samples. Clear thinking about the statistical use of samples can pay off by helping us understand better other kinds of things we might do with samples as well.

WHAT IS SAMPLING?

Sampling is the selection of a sample of elements from a larger *population* (sometimes called a *universe*) of elements for the purpose of making certain kinds of inferences about that larger population as a whole. The larger population, then, consists of the set of things we want to know about. This population could consist of all the archaeological sites in a region, all the house floors pertaining to a particular period, all the projectile points of a certain archaeological culture, all the debitage in a specific midden deposit, etc. In these four examples, the elements to be studied are sites, house floors, projectile points, and debitage, respectively. In order to learn about any of these populations, we might select a smaller sample of the elements of which they are composed. The key is that we wish to find out something about an entire population by studying only a sample from it.

WHY SAMPLE?

It at first seems to make sense to say that the best way to find out about a population of elements is to study the whole population. Whenever one makes inferences about a population on the basis of a sample there is some risk of error. Indeed, sampling is often treated as a second-best solution in this regard – we sample when we simply cannot study the entire population. Archaeologists are almost always in precisely this situation. If the population we are interested in consists of all the sites in a region, almost certainly some of the sites have been completely and irrevocably destroyed by more recent human activities or natural processes. This problem occurs not only at the regional scale – rare is the site where none of the earlier deposits have been destroyed or damaged by subsequent events. In this typical archaeological situation the entire population we might wish to study is simply not available for study. We are forced to make inferences about it on the basis of a smaller sample, and it does no good to just close our eyes and insist otherwise. The unavailability of entire populations for study raises some particularly vexing issues in archaeology, to which we will return later.

We might not be able to study even the entire available population because it would be prohibitively expensive, because it would take too much time, or for other reasons. One of the most interesting other reasons is that studying an element may destroy it. It might be interesting to contemplate submitting an entire population of, say, prehistoric corn cobs for radiocarbon dating, but we are unlikely to do so since afterward there would be no corn cobs for future study of other kinds. We might choose to date a sample of the corn cobs, however, in order to make inferences about the age of the population while reserving most of its elements for other sorts of study.

In cases where destructiveness of testing or limitations of resources, time, or availability interfere with our ability to study an entire population, it is fair to say that we are forced to sample. Precisely such conditions often apply in the real world, so it is common for archaeologists to approach sampling somewhat wistfully – wishing they could study the entire population but grudgingly accepting the inevitability of working with a sample. Perhaps the most common situation in which such a decision is familiar concerns determination of the sources of raw materials for the manufacture of ceramics or lithics. At least some techniques for making such identifications are well established, but they tend to be time consuming, costly, and/or destructive. So, while wishing to know the raw material sources for an entire population of artifacts, we often accept such knowledge for only a sample from the population.

Often, however, far from being forced to sample, we should choose to sample because we can find out more about a population from a sample than by studying the entire population. This paradox arises from the fact that samples can frequently be studied with considerably greater care and precision than entire populations can. The gain in knowledge from such careful study of a sample may far outweigh the risk of error in making inferences about the population based on a sample. This principle is widely recognized, for example, in census taking. Substantial errors are routinely recognized in censuses, resulting at least in part from the sheer magnitude of the counting task. When a population consists of millions and millions of elements it is simply not possible to treat the study of each element in the population with the same care taken with each element in a much smaller sample. As a consequence, national censuses regularly attempt to collect only minimal information about the entire population and much more detailed information about a much smaller sample. It is increasingly common for the minimal information collected by a census of the entire population to be "corrected" on the basis of a more careful study of a smaller sample, although legislators may be opposed (either because they just don't understand the principles or because the corrections would be to the political advantage of their opponents).

Archaeologists are frequently in a similar position. Certain artifact or ecofact categories from even a modest-sized excavation may well number far into the thousands. Detailed characterization of lithic debitage, for example, can be a very time-consuming process. We are likely to learn considerably more from a detailed study of a sample of lithic debitage than from a cursory study of an entire population of thousands of waste flakes. In such situations we should eagerly embrace effective sampling techniques as an improvement over the study of entire populations.

HOW DO WE SAMPLE?

If the purpose of sampling is to make inferences about a population on the basis of a smaller sample of elements selected from it, then it is important to select the sample in such a way as to maximize the chance that it accurately represents the population from which it is selected. *Random sampling* is a very effective way to maximize this chance of accuracy, and we select samples randomly whenever possible.

We are all familiar with many ways to select samples randomly. The practices of drawing straws, drawing names from a hat, and turning containers round and round to spill out bingo numbers are all efforts at random selection. Such physical methods seldom achieve true randomness, but a proliferation of governmental lotteries has spawned a multitude of mechanical contraptions that select at least very nearly truly random numbers (all in a manner designed to be engaging to the home audience watching the drawing on television).

Perhaps the most common means of selecting a random sample is to number each element in the population from which the sample is to be drawn (from 1 to however many there are in the population). A list of random numbers then identifies the elements that will make up the sample. The list of random numbers may come from a computer program or it may come from a random number table like Table 7.1.

Suppose you want a list of ten random two-digit numbers (that is, numbers between 00 and 99). Pick a number in the table to use as a starting point by closing your eyes and stabbing your finger at the table. Your finger may land on the number 51 that appears in the third column of the fifth row. You could read off the next ten numbers across the fifth row so that your ten random numbers would be 63, 43, 65, 96, 06, 63, 89, 93, 36, and 02. Or you could read down the third column for 34, 76, 59, 42, 82, 27, 23, 27, 38, and 95. Or you could read to the left on the fifth row, for 50 and 51, and then drop down to the sixth row and read back across toward the right to continue with 96, 65, 34, 00, 41, 60, 29, and 64. You can read in either direction column-wise or row-wise from the starting point you select.

The principal rule for proper use of a random number table is never to use it in exactly the same way twice. If you need a second sequence of random numbers, close your eyes again and pick a new starting point, or read in a different direction than you did the previous time (or both). You just do not want to use the same sequence of random numbers over and over again; make a fresh start each time you select a sample. (And watch out for calculators that claim to generate random numbers. Some of them simply generate the same sequence of "random" numbers every time you press the button.)

Table 7.1. Random Numbers

50 79 13	18 85 26	80 01 74	73 44 03	81 25 58	14 74 59	91 56 48	88 67 99	04 91 80
17 97 55	39 91 18	43 28 73	68 74 25	62 87 14	53 69 21	35 22 37	12 45 85	14 74 75
38 48 77	82 81 82	47 75 62	63 44 62	38 12 64	22 93 81	52 10 62	45 07 53	74 39 93
76 87 58	73 88 35	35 16 46	31 38 60	51 36 31	55 34 69	09 34 67	60 31 73	10 37 43
51 50 51	63 43 65	96 06 63	89 93 36	02 25 02	47 75 46	02 50 01	72 55 10	56 69 09
96 65 34	00 41 60	29 64 23	61 71 94	61 38 48	70 10 91	48 83 73	02 93 32	08 69 07
91 22 76	00 63 04	07 14 17	18 60 19	11 75 72	86 97 67	69 98 09	11 98 17	52 99 69
28 99 59	78 92 33	29 54 62	17 78 29	57 52 54	74 64 14	20 47 00	94 97 43	46 33 07
81 53 42	15 05 38	14 09 83	44 66 04	06 10 42	14 28 62	75 62 28	49 00 75	52 48 09
32 95 82	45 22 67	42 78 47	47 19 89	18 84 62	24 49 82	40 00 97	99 13 75	46 75 18
59 25 27	06 30 60	19 87 34	27 10 04	94 28 21	59 82 96	16 68 69	74 36 58	19 90 19
01 41 23	34 37 75	30 24 21	41 34 04	18 18 74	66 91 46	27 09 99	91 20 19	33 59 60
34 58 27	03 62 01	58 59 98	01 86 10	12 08 74	52 23 66	42 85 72	02 49 45	22 60 68
61 33 38	19 16 16	71 71 61	23 70 21	57 63 95	14 91 04	47 37 98	26 77 37	95 34 20
91 75 95	57 13 78	90 20 21	42 56 54	36 71 43	42 17 99	06 54 58	81 33 64	92 26 61
40 66 19	64 53 15	27 39 11	28 71 36	65 70 23	34 43 27	89 67 31	31 12 85	80 73 35
80 55 13	01 99 94	72 29 87	73 06 68	87 97 33	27 62 51	52 33 17	72 90 06	72 37 11
45 87 71	15 94 31	09 98 88	64 20 05	11 84 10	14 91 15	80 68 26	56 03 22	10 08 18
19 30 96	02 25 42	68 26 34	79 50 41	64 32 71	90 43 20	91 68 04	07 38 05	30 34 26
60 38 33	50 59 24	73 82 64	65 28 09	32 04 76	63 81 96	83 68 90	52 43 68	89 44 57
22 94 75	27 41 32	86 21 91	49 13 71	57 56 28	12 40 56	03 54 54	47 92 27	29 18 91
25 23 23	20 26 36	48 13 17	54 42 97	63 86 42	64 65 01	69 49 32	87 79 24	49 96 79
59 51 80	91 35 81	29 17 19	19 71 29	76 87 03	97 67 52	21 47 29	20 01 39	33 37 45
05 40 65	66 23 54	23 94 43	44 09 08	81 12 79	58 01 74	81 60 89	70 89 43	37 53 90
61 99 79	13 20 09	56 58 07	59 70 46	32 86 47	36 81 20	89 89 98	71 94 37	88 72 58
24 34 19	08 05 18	51 49 14	30 48 09	47 94 63	12 04 80	76 38 53	09 37 03	04 06 53
29 48 01	18 37 83	94 16 20	37 09 53	63 72 89	96 74 35	13 21 80	77 54 24	09 72 15
65 78 94	61 74 72	11 71 52	15 71 62	98 87 73	39 41 82	12 98 31	83 67 01	86 03 52
04 24 77	46 63 39	03 10 85	10 79 39	08 17 74	64 84 20	43 21 22	46 26 73	51 41 17
73 71 88	69 64 06	08 26 63	51 35 45	66 52 78	38 85 11	80 39 30	86 85 48	44 46 43
88 59 20	63 92 58	52 12 02	37 13 31	42 52 34	77 50 18	09 17 48	46 41 32	83 26 01
84 82 52	27 55 25	20 16 11	66 94 25	04 94 55	79 03 65	61 21 49	97 72 46	56 26 52
82 26 26	52 50 21	63 86 14	11 69 21	98 97 03	68 59 09	98 34 50	58 38 79	03 64 69
81 52 82	82 86 08	45 99 54	14 71 46	14 01 68	33 59 29	71 09 23	37 84 04	92 61 34
90 95 02	61 36 94	98 81 54	90 60 64	84 49 23	92 30 99	69 65 65	47 54 73	17 81 21
37 78 13	13 55 40	07 53 92	98 82 64	01 11 08	94 91 84	83 55 46	30 96 74	13 54 30
01 87 88	82 01 76	59 28 87	03 73 69	22 99 27	30 62 73	02 34 82	30 59 37	27 95 50
02 96 02	54 62 25	36 56 61	38 80 15	93 30 11	34 67 53	81 83 54	83 86 47	64 43 03
40 53 25	64 31 38	89 14 23	54 33 86	58 03 94	57 03 68	78 38 14	20 09 42	82 84 06
46 81 46	18 47 75	70 20 70	33 15 43	73 67 61	05 55 50	03 15 86	55 91 52	73 90 95
69 72 68	17 87 22	62 08 49	40 32 38	25 71 59	29 67 81	23 68 36	49 94 65	15 03 72
26 24 90	53 49 35	91 07 60	74 61 62	06 07 67	95 99 56	28 56 02	52 61 94	81 14 33
68 17 38	10 48 60	81 73 25	34 55 76	40 84 05	23 55 96	20 60 74	08 03 42	51 81 07
06 51 06	07 44 30	86 12 69	99 16 51	10 05 54	16 07 18	16 24 26	09 97 30	57 50 11
45 52 21	16 03 36	28 32 27	25 44 46	14 17 81	29 86 97	59 12 03	67 28 83	33 03 64
54 72 12	20 91 87	53 87 29	39 84 26	59 80 66	44 84 84	63 77 81	31 48 92	45 99 33
72 65 08	37 37 55	91 23 02	22 51 88	94 32 45	09 14 81	31 14 27	26 61 93	41 52 08
47 20 65	40 51 39	78 88 88	71 45 86	03 08 99	61 16 56	47 08 54	89 79 29	24 91 42
94 79 42	62 56 17	34 45 56	84 96 09	56 22 13	14 87 21	97 66 60	48 64 56	41 45 92
40 03 28	30 16 77	79 10 05	94 90 35	08 03 11	91 56 83	42 23 20	08 44 82	13 47 70

If you need one-digit numbers, you can treat each individual column of digits as a separate column in the table. If you need four-digit numbers, you can treat pairs of columns together as four-digit numbers. If you need three-digit numbers, you can simply ignore the first or last digit in a four-digit number. The spaces dividing the numbers into columns of two-digit numbers, in short, are entirely arbitrary, as are the wider spaces grouping the columns by threes. Likewise, the extra space setting off groups of five rows each is included only to make the table easier to read.

Suppose the population from which you wish to select a sample contains 536 elements, numbered from 001 to 536. You need a list of three-digit random numbers between 001 and 536. You can select a list of numbers in exactly the manner just described, except that you ignore any number less than 001 (that is, 000) or any number greater than 536 (that is, 537–999). You simply skip past these inapplicable numbers in the list and continue to select those in the relevant range until you have as many as needed.

Sometimes the same number will appear more than once in a list of random numbers. If this happens, you can follow either of two courses. The first is to ignore multiple appearances of the same number and continue reading the random number table until you have as many numbers as you need without repetitions. This is called sampling *without replacement*. (The name makes sense if you imagine that you were actually drawing numbered slips from a hat without replacing the slips in the hat for potential re-selection on subsequent drawings.) Sampling without replacement is the course of action that seems to make intuitive good sense to many people.

Sampling with replacement, however, turns out to be a little simpler mathematically, and the equations in this book are those for sampling with replacement. In sampling with replacement, each time you draw a numbered slip from the hat (speaking metaphorically), you write down the number and replace the slip in the hat so that it could be drawn again in the future. The analogous procedure, when sampling with a random number table, is to include repeated numbers in the sample as many times as they appear in the list from the random number table. The data for the corresponding elements, then, are included among the sample data as if each occurrence in the sample were an entirely different element.

Suppose, for example, that we were sampling with replacement from a population of scrapers, in an effort to estimate the mean length of scrapers in the population. The random numbers chosen from the table might be 23, 42, 13, 23, and 06. We would select the scrapers with the numbers 06, 13, 23, and 42 and measure the length of each. We would, however, write down five length measurements, not four, so as to include the length measurement for scraper number 23 twice. The number of elements in the sample would also be five, not four. To re-emphasize, *it is this procedure, sampling with replacement, that the equations in this book are appropriate for.* Slightly different equations are technically necessary for sampling without replacement, although in almost every practical instance it makes very little meaningful difference in the results. It is, however, quite easy to adhere strictly to the assumptions on which the formulas given in this book are based simply by including the data from an element in the sample as many times as that element is selected.

REPRESENTATIVENESS

The kind of sample selection we have just discussed is called *simple random sampling*. The effect of using a table of random numbers is to give each individual element in the population an equal chance of selection, and this is the most straightforward way to phrase the essential principle of simple random sampling. It is because each individual element in the population has an equal chance of inclusion in the sample that a random sample provides us with our best chance of obtaining a sample that accurately represents the population.

The concept of *representativeness* is a slippery one, worth discussing more fully. As noted at the beginning of this chapter, our aim in sampling is to make inferences about a population of elements based on a sample from that population. We take that sample to represent the population, so the representativeness of the sample is of critical importance. The problem is that, without studying the entire population, we can never be absolutely certain that the sample represents it accurately. If we intend to study the entire population, of course, there is no need to worry about the representativeness of a sample. It is only if we do not intend to study the entire population that we must worry about the representativeness of a sample, but that is precisely the situation in which we cannot provide any guarantee of representativeness. It is this difficulty that much of the rest of this book is about. (Statistics books have more in common with Joseph Heller's *Catch 22* than is often noticed.)

Some archaeologists seem to have the impression that if a sample has been selected randomly, then it is guaranteed to be representative. Nothing could be farther from the truth. Like samples not selected randomly, random samples represent the populations from which they were selected sometimes quite accurately, sometimes with moderate accuracy, and sometimes very inaccurately. Random sampling, while it does not provide a guarantee, does give us our best chance at a representative sample. *Most important of all, random sampling provides a basis for estimating how likely it is that our inferences about the population are wrong, and thus tells us how much confidence we should place in these inferences.*

DIFFERENT KINDS OF SAMPLING AND BIAS

Simple random sampling is, as its name implies, the simplest and most straightforward method of selecting a random sample. In most of the rest of this book, mention of random samples refers to simple random samples. There are other somewhat more complicated variants of random sampling. They are best dealt with after the implications of simple random sampling are fully explored and understood, so we will not discuss them in any detail here. It is important to recognize their existence at this point, though, in order to understand the limits of applicability of the methods appropriate to simple random sampling that are the subjects of the following chapters.

When the population we wish to make inferences about can be readily divided into different subpopulations, it is often advantageous to select subsamples separately from each subpopulation. It may be the case that some subpopulations are more intensively sampled than others. If so, an element in a subpopulation that is more intensively sampled has a greater chance of inclusion in the overall sample than an element in a subpopulation that is less intensively sampled. This violates the fundamental principle of simple random sampling. When we select separate random subsamples from different subpopulations, we call it *stratified random sampling*. An instance in which we might apply stratified random sampling would be in selecting samples of ceramic sherds for raw material sourcing from each of the eight known households at an excavated site. Each household would be a *sampling stratum*, and we would make inferences about where the ceramics in each household were made based on independent samples. All eight samples might later be combined for purposes of making inferences about where ceramics at the settlement as a whole were made. Stratified random sampling is discussed in Chapter 16.

When the elements of the population we wish to make inferences about are not available individually for selection, we often use sampling strategies based on spatially defined selection units. If the population we are interested in, for example, consists of the lithic artifacts in a particular site, we can likely select a simple random sample of artifacts only if the site has already been excavated and the artifacts are in a laboratory or museum where we can select sample members individually. If the site has not been excavated we may nonetheless wish to obtain a random sample of lithic artifacts from it.

This could be done by excavating small test pits in a number of different locations to recover some of the artifacts that lie buried in the site's deposits. If the locations of those test pits were randomly selected (for example, by establishing a grid system over the site area and randomly selecting which grid units to excavate), then the resulting artifacts would still be a random sample of artifacts from the site. They would not, however, be a simple random sample because the elements in the sample (that is, lithic artifacts) were not individually selected. It was grid units that were individually selected randomly, and so we do have a simple random sample of grid units. But it is a population of lithic artifacts, not a population of grid units that we wish to make inferences about in the present instance. Lithic artifacts were selected, not individually, but rather in small groups or *clusters*, each cluster being those lithic artifacts contained in the deposits in one excavated grid unit. We then have, not a simple random sample of artifacts, but rather a *cluster random sample* of artifacts. Cluster samples, like stratified samples, are within the reach of statistical tools – the ones taken up in Chapter 17.

Several other terms have come to be used here and there in the archaeological literature for nonrandom ways of selecting samples – "haphazard sampling," "grab sampling," "judgmental sampling," "purposive sampling," and the like. These are not well-established terms that have clear meanings with precise statistical implications. They refer to the explicit or implicit application of a variety of nonrandom selection criteria. In some circumstances it may be justifiable to treat such samples as if they were random samples, but such treatment must be applied with caution, and specific justification for it is always required.

For example, a surface collection made at an archaeological site is sometimes described as a haphazard sample or a grab sample, probably meaning that field workers walked around an area and picked up haphazardly an assortment of the things they saw. The resulting sample is then sometimes used as a basis for making inferences about the population from which this haphazard sample was selected, presumably the entire set of artifacts on the surface of the site at the time. This approach is likely to produce a sample that systematically misrepresents the population in certain respects. For example, a haphazard surface collection of this sort is likely to contain a higher proportion of large artifacts than the population did, simply because they will be more noticeable. As a consequence, if the sample is used to make inferences about the average size of artifacts on the surface of the site, the inferences will be inaccurate. Similarly, if the sample is used to make inferences about the proportions of different artifact classes on the surface of the site, classes of artifacts that tend to be large will seem to be more abundant than they really were.

Much the same could be said about color and other characteristics that affect the visibility of artifacts lying on the ground. Even more subtly, a haphazard surface collection like this may have higher proportions of unusual things and correspondingly lower proportions of common things than does the artifact population from which it comes, as a consequence of a subconscious tendency to collect things that strike the eye as especially different from most of the other things being seen.

This haphazard sample is *biased* because the elements in it were selected in a way that makes the sample systematically different from the population in certain respects. There is no statistical technique for eliminating such bias once the sample has been selected. The appropriate statistical tool for avoiding sample bias is random selection of the sample, and this tool must be used at the time the sample is selected. It cannot be applied retroactively. Haphazard or grab samples are simply not the same as random samples.

Judgmental or purposive samples are also likely to be biased. These terms tend to refer to samples selected by looking over the range of elements in a population and specifically deciding to include certain elements in the sample and exclude others. Obviously, whatever the criteria involved in the selection, the resulting sample will be biased with respect to those characteristics. Suppose, for example, that an archaeologist wishes to study the residential remains at a site where the locations of individual households are marked on the surface by small mounds. He or she might decide to thoroughly excavate those mounds that show the highest densities of artifacts on their surfaces on the theory that their excavation will produce greater numbers of artifacts. The result, of course, will be a sample of house mounds with a substantially higher number of artifacts than average for the site as a whole. Such a sample could clearly not be used to make inferences about the average density of artifacts in house mounds at the site.

Still more insidious are other possibly related factors. The higher artifact densities that caused mounds to be included in the sample might be the result of, say, the greater wealth of certain households and consequently more intensive disposal of used and broken objects near them. Thus the sample might systematically misrepresent the wealth of households at the site and therefore be biased in this regard as well. Inferences about the entire population in regard to such characteristics as the proportions of different artifact or ecofact classes related to wealth would be systematically erroneous.

Once again, the moral of the story is that random selection of the elements in a sample is the only way to ensure that a sample is unbiased. Random samples are the only ones that we can be sure are unbiased, because their method of selection specifically avoids conscious and subconscious biases of the kinds just discussed. Since bias refers to the systematic application of criteria that result in an unrepresentative sample, we know that biased samples are unrepresentative in certain ways. The reverse is not, however, true. That is, the absence of bias from random samples does not guarantee that each and every one will accurately represent its parent population. We can never be entirely certain that a sample accurately represents the population from which it was selected (unless we study the entire population). Biased samples are known to be unrepresentative in some ways, but we do *not* know that unbiased (that is, random) samples are "representative." An archaeologist who selects a sample randomly so as to know that it is "representative," labors under a serious misunderstanding of sampling principles. We can (and will in the next few chapters) assess the probability that a random sample is unrepresentative, something we cannot do except with random samples.

USE OF NONRANDOM SAMPLES

Most of the statistical tools discussed in this book require us to assume that we are working with random samples. Most archaeological data, however, were (and still continue to be) produced with nonrandom sampling procedures that result in biased samples. This, on the surface, would suggest that statistical tools are not applicable to most archaeological data. And this, indeed, is the conclusion at which some archaeologists have arrived. The situation, however, is simultaneously both more and less serious than this.

First the bad news. The difficulty of making inferences about populations from samples we know to be biased is not unique to statistical means of making inferences. We cannot reliably estimate the average size of artifacts on the surface of a site from a collection that over-represents large artifacts. This is true whether our means of inference are statistical or purely intuitive. We are simply unable to make conclusions by any means about the average size of artifacts in the population on the basis of such a sample. It is no solution to avoid statistical approaches and rely strictly on subjective impressions or any other kind of inference, because all kinds of inferences are affected in precisely the same way by sampling bias. Thus, the need for unbiased samples does not derive from arcane rules of statistical inference. It is fundamental to inference of any kind, and we ignore it only at our own peril, whether we use statistical tools or not.

Now the good news. This is not another cautionary tale ending at the nihilistic conclusion that reliable interpretation of archaeological remains is impossible. Too many such tales have already appeared in the archaeological literature. Thorough understanding of the nature of sample bias and careful application of common sense can make inferences about populations from samples possible. Moreover, clear thinking about this issue, stimulated by efforts to apply statistical techniques, can be carried over productively into the arena of nonstatistical inference making, leaving us in position to make more reliable conclusions in other ways as well.

The effects of known or possible bias in sample selection can (and must) be evaluated with particular reference to at least three specific ways in which biased samples might still be used.

First, a sample that is biased in one respect is not necessarily useless for any and all purposes since it may not be biased in other respects. If a case can be made that the bias in sample selection is unrelated to some other characteristic of the population, then the sample might be appropriate for making inferences about that other characteristic.

Second, two samples selected with the same bias may still be usefully compared even with regard to the characteristic involved in the selection bias. Here the case that must be made is that the bias operated similarly enough in the selection of both samples to have had a very similar impact on both. The two samples then might be unrepresentative of their parent populations in precisely the same way, making some kinds of conclusions from comparing them reliable.

Third, in some instances, useful comparisons may be made, even between samples selected with different biases related to the characteristic of interest. If a sample from one population is selected with a bias in favor of some characteristic while a sample from another population is selected with a bias against it, that characteristic should be more abundant in the former sample. If comparing the two reveals that the characteristic is actually *less* abundant in the former sample, this cannot be a consequence of sampling bias, which would produce the opposite effect. This outcome would sustain an inference that the population the former sample came from had more of the characteristic of interest than the population the second sample came from. The difference between the populations in this regard could be argued to be even stronger than the difference observed between the samples.

Judgments in instances like these involve ad hoc reasoning more than the application of general rules or principles, and the process is, perhaps, made clearest through examples rather than abstract discussion.

Example: A Haphazard Surface Collection

In the instance of a haphazard collection of artifacts from the surface of a site, very small artifacts are almost certainly not collected as frequently as larger ones are, simply because they are considerably less noticeable. (If we want to be truly honest about it, the same could surely be said of most artifact samples recovered from screens during excavation.)

In the instance of a surface collection, we probably have no interest in inferring anything about the average size of artifacts. It may be of considerable interest, however, to estimate the proportions of different ceramic types in the parent population. If we can make the case that sherd size is unrelated to ceramic type, then even a sample selected with bias in regard to sherd size can be used to make such inferences reliably. Only if some ceramic types tended systematically to break into substantially smaller pieces than others and therefore systematically to be under-represented in the sample would sampling bias on account of size affect inferences about the proportions of different ceramic types. The possibility of a relationship between sherd size and ceramic type could be evaluated empirically before pro-ceeding to use a haphazard artifact collection as the basis for such an inference. Similarly, other possible kinds of bias resulting from haphazard sample selection can be enumerated and their impacts on particular kinds of inferences that we are interested in assessed.

Even if we determine that sample bias makes our inference about proportions of different ceramic types suspect in this instance, this suspect inference might still be usefully compared with similar inferences concerning other sites based on sam-ples selected with the same biases. As long as the operation and strength of the bias in sample selection can be supposed to be the same for all the samples, then the inaccurate inferences may be, in effect, comparably inaccurate. A sample that under-represents a particular type can be usefully compared to another sample that under-represents the same type to the same extent. Comparisons are quite often the ultimate objective in working with type proportions anyway. It likely has no particu-lar meaning to us that a particular type comprises, say, exactly 30% of the ceramics on the surface at some site – only that this 30% is greater than the figure of 15% obtained from another site. It usually makes no difference at all, finally, that the truly accurate numbers might be 36% and 18%, respectively, instead of 30% and 15%. For comparative purposes, then, sampling bias may, in effect, cancel itself out when it affects all samples in the same way and to the same extent.

Even when biases are very different from one sample to another, some com-parative conclusions might be drawn. Suppose a haphazard surface collection is made carefully by archaeologists attempting to pick up all the artifacts they can see, and 8% of the artifacts in the collection turn out to be fragments of small ceramic figurines. No matter how careful the archaeologists were, this collection probably contains some bias in favor of figurine fragments because their unusual shapes make them easier to spot than many other artifacts. The proportion of figurine fragments in the entire population of artifacts on the surface of the site is probably actually some-what less than 8%. Another site is visited by archaeologists much less concerned about sampling bias who casually pick up several bags of whatever artifacts hap-pened to attract their attention. We would suspect a considerably stronger bias in favor of figurine fragments in this latter collection, but the proportion of figurine fragments here turns out to be 3%. The proportion of figurine fragments in the entire population of artifacts on the surface of this second site is probably actually substantially below 3%.

With this outcome, it is extremely likely that the first site really does have a larger proportion of figurine fragments on its surface than the second does. The biases operating in selecting samples (which is what the surface collecting really amounts to) were not the same. If the two sites actually had the same proportion of figurine fragments on their surface, surely the second sample would have contained a higher proportion of them than the first. But the second sample actually contained a *lower* proportion – a difference between the two samples that could not possibly have been produced by the sampling bias, which would have had the opposite effect. The higher proportion of figurine fragments in the first sample must have been produced by a real difference between the two sites with regard to the proportions of figurine fragments on their surfaces, and this difference is surely even greater than the difference between 8% and 3%. In other words, "somewhat less than 8%" (from above) differs from "substantially below 3%" (also from above) by even more than the 5% that separates 8% from 3%.

The ability to make comparative inferences about populations in this last instance depends on the outcome. If the second collection had contained 15% figurine fragments, we would not be able to conclude much. It would be entirely possible that the difference between the two samples was the result of nothing more than the stronger sampling bias in favor of figurine fragments in the second sample. In order to arrive at the point of making the conclusions that might be made in these circumstances, one has to be willing to set aside (temporarily) worries about sampling bias, go ahead and compare the percentages, and then think again about sampling bias in light of the results. Sometimes that final thinking will lead to the conclusion that the comparison tells us nothing reliable; at other times it may lead to conclusions we can make with considerable confidence.

Example: A Purposive Obsidian Sample

Many archaeologists have been faced with a sampling decision in regard to raw material sourcing. Obsidian artifacts, for example, from many parts of the world can be linked to sources of raw material through chemical fingerprinting. The necessary analyses, however, are so expensive that it is usually possible to identify only a portion of the obsidian recovered from a site. In this situation some archaeologists have looked over all the obsidian obtained from the site and selected as many pieces as they can afford to analyze, intentionally including artifacts of as many different colors and appearances as possible. The justification for this procedure has usually been that it provides the greatest chance of including material from the largest possible number of different sources since material from different sources may differ visually as well as chemically. Since some sources may be represented by only a few pieces in a large population, there is a very good chance that those sources might not turn up at all in a random sample of modest size – hence the interest in including in the sample for analysis pieces of very unusual appearance.

The sample selection is thus biased, systematically over-representing in the sample artifacts of unusual appearance. If there is, indeed, some relation between appearance and source location, this bias makes the sample irretrievably inappropriate for making actual estimates of the proportions of the artifacts that were made with materials from the different sources. To see why, one can imagine drawing a sample of 4 marbles from a jar with 97 black marbles, 1 blue marble, 1 red marble, and 1 green marble. Clearly the best representation of the full range of different colors in the population can be achieved by purposely selecting a sample consisting of one marble of each color. That sample, however, could not then be used to estimate the proportions of different colored marbles in the jar. Observing that the sample consisted of 25% black, 25% blue, 25% red, and 25% green would lead directly to the inference that the proportions in the population were also 25% of each color, but we know the real proportions to be 97% black, 1% blue, 1% red, and 1% green. The proportions in the sample were determined not really by any characteristic of the population but rather entirely by the biased sampling procedure.

A sample of obsidian artifacts selected in such a way for source analysis simply cannot be used to make inferences about the proportions of material acquired from different sources, no matter how useful it may be for obtaining as long a list as possible of different sources exploited. Comparisons with samples from other sites selected according to similar principles are impossible, since even the directions of the biases introduced cannot be guessed at. They would favor little-utilized sources, but we would never know which those were. Such a sample might be used for other inferences insofar as it is possible to argue that the bias in sample selection does not relate to these other inferences. For example, the sample might be used to study whether material from different sources tended to be worked in different ways. The selection bias would, superficially at least, not appear to relate to this issue.

Conclusions on Bias

The only way to be absolutely certain that sampling bias has no effect on inferences made, of course, is to be certain that sample selection is entirely free from bias. Random sampling is the appropriate technique for avoiding bias in sample selection and should be applied whenever possible (even when statistical means of making inferences are not contemplated). To the extent that the case can be made, however, that bias in the selection of already existing samples does not affect the specific inferences being made, then we can use those samples. And that means quite literally *to the extent that the case can be made* – to that extent; no more and no less. Like so many other things in life, sampling bias is not a matter of black or white but of varying shades of gray. Clear, careful thinking may convince us that the risk of sample bias, insofar as a particular inference is concerned, is minimal, even though the sample at hand is egregiously biased in other regards. If we can postulate a number of specific ways that bias might affect the inferences we are interested in and then empirically rule out all these possibilities, then the case for disregarding bias (and treating an existing sample as if it were a random sample for a particular purpose) may be quite convincing. If, in contrast, we simply ignore the possibility of such problems, then any inferences made must be viewed with suspicion.

This is not a perspective on sampling bias that is often expressed in the archaeological literature or elsewhere, and it certainly runs counter to the rules laid down in many statistics textbooks – particularly those of the cookbook persuasion. Statistics books that emphasize memorizing rules (the "Ours not to reason why" approach) are likely to forbid the application of most of the techniques discussed in this book to any sample not strictly randomly selected (by which they mean with a table of random numbers or similar procedure to preclude bias). This would mean that these techniques could not be applied to the vast majority of archaeological data now in existence. Worse yet, since, as we have seen above, sampling bias affects not just statistical inferences but any kind of inferences from samples to populations, we would not be in position to make any inferences at all from these data.

Most of those who adopt such a stringent position will probably not be much attracted to archaeology, and will not be reading this book. Archaeologists who do see things this way will continue to write cautionary tales emphasizing that we can make few, if any, interesting conclusions from archaeological information. The rest of us will just continue to do the best we can with what the archaeological record provides. (It has been pointed out by others that archaeology falls not among the hard sciences, but instead among the difficult sciences.) This last group should take advantage of the proper techniques of random sample selection to guard against sampling bias to the maximum extent possible. When we cannot do this (as, for example, to learn what can be learned from previously selected samples), we must use our wits to assess the nature and strength of the impact sampling bias may have on particular inferences. Sometimes we will make inferences that must be taken with caution because some impact from sampling bias cannot be ruled out. If these inferences prove interesting, they may justify further data collection to see if they hold up even with unbiased samples.

THE TARGET POPULATION

The previous discussion may imply that adoption of strict random sampling procedures could resolve the issue of sampling bias in archaeology once and for all by avoiding it entirely. An even stronger view, sometimes argued in the archaeological literature, is that sampling bias is best avoided by abandoning sampling altogether in favor of studying entire populations. Neither of these solutions, however, will work in archaeology because the *target population* that we wish to make inferences about is seldom fully available either to study in its entirety or to select a sample from.

At a regional scale, at least some sites in any region are likely to be unavailable for study because they have been covered by modern urban concentrations, obscured by recent sedimentation, carried away by erosion, or otherwise destroyed or made inaccessible. Thus even a regional survey that is complete in the sense of systematically covering the entire surface of the region does not have access to the complete population of sites that we need to study. A sample selected by the strictest random sampling procedures remains, not a sample of all the archaeological sites ever left, but a sample of those sites that remain accessible for selection. Similarly, at a smaller scale, the vast majority of archaeological sites are not intact and completely preserved but are only partial, with some sectors destroyed or inaccessible to study. Thus, whether we study entire archaeological populations or random sample them, the populations truly available for study or sampling do not precisely correspond to the populations we wish to make inferences about.

Random sampling puts us in position to make inferences about the population the sample was drawn from, and, of course, study of an available population provides us with conclusions about that available population. If the available population, however, was only the part of an important site that had not been washed away by the adjacent river, we are faced with the difficult question of how to attempt to characterize the entire meaningful site. There is no simple and straightforward solution to this difficulty, just as there is no simple and straightforward solution to the problem of making inferences from biased samples. The most common response by archaeologists to this difficulty is simply to ignore it. This response is clearly conceptually inadequate, although a number of famous archaeologists have built successful careers on it. Another common response is to pretend that the missing part of the site contained what we hoped to find but didn't find in the part that remains. This is just plain unconvincing.

Fundamentally the difficulty of not being able to study or sample from the population we are truly interested in parallels the problem of sampling bias. The population available to be studied or sampled is, in effect, itself a sample from the target population – one selected by quite possibly very biased procedures (whatever processes destroyed or made inaccessible the portion we cannot now study). It is because archaeologists are so frequently in this position that they are forced to sample in one way or another. Often the entire population available for study is already a sample. We thus cannot escape the complexities of sampling and the issue of sampling bias, no matter how we try.

Whether we select samples ourselves, work with data from samples other people selected, or study entire available populations, we still must wrestle the sampling bias problem to ground as best we can if we propose to do archaeology at all. This means using our understandings of sampling and sampling bias to say as much about the representativeness of a sample as possible, using statistical tools presented in this book and/or using nonstatistical and probably ad hoc reasoning applicable to specific instances.

Even when we can apply random sampling procedures to an available population that corresponds well to the target population about which we wish to make inferences, avoiding sampling bias with random selection does not guarantee a representative sample, as discussed above. It only gives us the best chance of a representative sample and enables us to assess the probabilities of its unrepresentativeness.

In any of these cases, then, some of the inferences we make about entire populations from the samples we can study will be correct and some will be incorrect. Some will be incorrect because the population from which we could select a sample did not represent very accurately the population about which we wish to know. Some will be incorrect because the sample we study does not accurately represent the population from which it was selected. Although related, these are two different sources of error. The first must be dealt with on an ad hoc basis with cleverness and common sense. Random sampling and the statistical tools discussed in the next few chapters can help us with the second by telling us roughly what percentage of our inferences are incorrect for this reason. We cannot, however, determine specifically which ones are incorrect. Without these tools we can say even less. If we are careful and diligent, most of our conclusions will be correct, but it is unrealistic to hope to make correct inferences 100% of the time, no matter how careful we are to eliminate sampling bias (and other inaccuracies). Finally, confidence in our ultimate conclusions is best reinforced by finding consistent patterns in the majority of multiple independent inferences. When such consistent patterns are recognized, it should make us willing to set aside inconsistent inferences as possible consequences of sampling error (of either of the two kinds mentioned above).

To those who are concerned that I have taken here too cavalier an attitude toward the importance of random sampling in statistical (or other) inference, I can only say that I see no other way to proceed in most of the situations that practicing archaeologists must actually face. The course advocated here is to try to rule out all the likely ways in which a sample may be biased. If it seems likely that a sample may be unbiased, then it is worth setting our quite proper worries about sampling bias aside for the moment at least and going ahead to see what inferences about the population our sample may lead to. If it does lead to interesting inferences about the population, then our worries about sampling bias must return as the proverbial grains of salt with which our conclusions are taken. If we are fairly confident that the sample we are working with can be taken to be unbiased, then we can be fairly confident about the conclusions concerning a population that we make on the basis of that sample. If we think the sample we are working with might be biased, then whatever conclusion we arrive at about a population on the basis of that sample must be taken with a correspondingly large grain of salt.

Practitioners of most other disciplines do not find these issues as troublesome as archaeologists do, because they are usually interested in studying target populations that are much more accessible for study than those of the archaeologist. They can often afford to ignore results from samples that may be biased and simply go back to the field or laboratory for a more carefully selected sample. Much of the sampling bias in archaeological samples, however, is not so easily avoided. We must learn, then, to avoid sample bias whenever we can (as by selecting truly random samples) and to live and work productively alongside it when we must. When the discrepancies between our real target population and the population actually available to be sampled are truly large, excessive finickiness about sample selection procedures begins to be like straightening deck chairs on the Titanic. We need to be careful and thoughtful in deciding when straightening deck chairs is a worthwhile activity and when our attention would better be directed to the lifeboats.

Much of this discussion anticipates the statistical techniques discussed in Chapters 8–10 and may not make too much sense to those who do not already have some inkling of them. The issues raised will come up again repeatedly in this book, however, and the discussion in this chapter lays out the reasons for approaching them in the way that we will. We will return to them in the last chapter as well.

PRACTICE

1. Imagine that you have made an intensive surface collection at the Keeney Knob site. The following Saturday night you happen to meet someone who used to own a farm at Stony Point. Later on, he lets you study the large collection of lithic artifacts he made on his farm before they built the shopping center and obliterated all trace of the archaeological site. You immediately recognize that the lithics from Stony Point are precisely contemporaneous with the ones you have collected at Keeney Knob, and you are eager to compare the artifacts from the two sites. First, you would like to know whether the Keeney Knob and Stony Point lithic assemblages have similar or different proportions of projectile points. Of the artifacts in your surface collection from Keeney Knob, 14% are projectile points; of the collection from Stony Point, 82% are projectile points. Second, you are interested in the raw materials from which projectile points were made at the two sites. Of the Keeney Knob projectile points, 23% are obsidian, and 77% are chert; of the Stony Point projectile points, 6% are obsidian, and 94% are chert. You recognize, however, that you have a potential problem of sampling bias in making use of these comparisons. How would you assess this problem and what would you do about it? Can you make any use at all of these comparisons? Can you be more confident about conclusions from one of them than from the other? Why?

2. You have data from haphazard surface collections at a series of neolithic sites in the Velika Morava River valley. They were made during a field season in 1964 by a research team, doing what experienced archaeologists usually do to sites, before the area was flooded by a reservoir, ending all possibility of further archaeological research. If your hypothesis about the beginnings of grain cultivation in the region is correct, the sites in river bank locations should have substantially larger proportions of stone hoes than the sites set back from the river. What worries about sampling bias would you have to face in using the data from the 1964 Velika Morava survey to investigate your hypothesis? How would you face these worries? How much confidence would you place in conclusions you arrived at about the proportions of stone hoes at different sites in this region, based on the 1964 survey? Why?

Chapter 8
Different Samples from the Same Population

The discussion in Chapter 7 dealt with the fact that sometimes random samples represent the populations from which they are drawn very accurately and sometimes they don't. Random selection is no guarantee of representativeness. Random sample selection does, however, make it possible to apply some very powerful tools for assessing how likely it is that a sample is unrepresentative to a particular degree. This is because, with the unbiased samples that random selection produces, we can say something about how often particular degrees of unrepresentativeness occur, on average.

ALL POSSIBLE SAMPLES OF A GIVEN SIZE

In order to understand this we must consider the many possible different random samples that can be drawn from a single population. Table 8.1 contains the measurements (in cm) of the diameters of 17 post holes from excavations at a single site. The measurements have been arranged in ascending order to make them easier to examine. We will consider these 17 measurements a population of measurements from which we wish to draw a sample. This is, of course, an exceedingly small-scale example. Samples themselves are likely to consist of far more than 17 measurements, and the populations from which the samples are drawn are even larger. But this small example enables us to see in operation principles that it would be almost impossible to observe in an example of large enough scale to be more realistic.

This population, then, consists of 17 post holes, whose diameters have been measured. We will use the capital letter N to indicate the number of elements in a population, thus, for this example $N = 17$. The mean post hole diameter for the 17 post holes in the population is 13.53 cm, and we will use μ (the Greek lower-case

R.D. Drennan, *Statistics for Archaeologists*, Interdisciplinary Contributions
to Archaeology, DOI 10.1007/978-1-4419-0413-3_8,
© Springer Science+Business Media, LLC 2004, 2009

Table 8.1. Diameter Measurements for a Small Population of Post holes[a]

Post hole number	Diameter (cm)	Post hole number	Diameter (cm)
1	10.4	10	13.2
2	10.7	11	13.7
3	11.1	12	14.0
4	11.5	13	14.3
5	11.6	14	15.0
6	11.7	15	16.4
7	12.2	16	18.4
8	12.6	17	20.3
9	12.9		

[a]$N = 17$; $\mu = 13.53\,cm$; $\sigma = 2.73\,cm$

letter mu) to stand for the mean of the population. Thus $\mu = 13.53$ cm. The standard deviation of a population is represented by σ (the Greek lower-case letter sigma), so in this example $\sigma = 2.73$ cm.

We will begin by considering the smallest possible sample, a sample of 1. The lower-case letter n represents the number of elements in a sample, just as N represents the number of elements in a population. We will consider all the possible samples of 1 ($n = 1$) that could be drawn from this population of 17 post holes ($N = 17$). It is easily seen that there are 17 possible different samples of 1 that might be selected. We might randomly select post hole No. 1, or No. 2, or No. 3, ... or No. 17. Whichever sample of 1 we happened to select, we could calculate the mean post hole diameter in that sample and use it to estimate the mean post hole diameter for the entire population. Our best guess for the population mean is always the sample mean. In order to distinguish these two means in equations we use μ to stand for the population mean as in Table 8.1 and $\overline{X}$ to stand for the sample mean. Thus, the best estimate of μ is always $\overline{X}$.

If our sample consisted of post hole No. 1, we would guess that the mean post hole diameter in the population was 10.4 cm, since 10.4 cm is the mean of a sample consisting of the single observation 10.4 cm. If our sample consisted of post hole No. 2, we would guess that the population mean was 10.7 cm, and so on. From the 17 different samples of 1 that we might select we could make 17 different guesses at the population mean. Some of these guesses would be very close (as for the samples consisting of post hole No. 10 or post hole No. 11). Other guesses would be much farther off (as for the samples consisting of post hole No. 1 or post hole No. 17). This example shows clearly that some samples represent the population from which they are drawn relatively accurately, and others do not.

The largest possible error in estimating the population mean occurs when the sample of 1 consists of post hole No. 17. On the basis of this sample we would guess that the population mean was 20.3 cm, an error of 6.77 cm. This is certainly a regrettably large error. Moreover, such a maximum error will occur fairly often in drawing samples of 1. Fully 1/17 (5.9%) of the total number of different samples of

1 that could be drawn from this population would consist of post hole No. 17. Thus, if we were to select samples of 1 repeatedly from this population, 5.9% of these samples would consist of post hole No. 17 and cause us to make such an erroneous guess at the population mean. Sampling in this way, then, we would make an error as large as 6.77 cm, 5.9% of the time.

If we needed to estimate the mean post hole diameter in this population with an error no greater than 3.0 cm, we could figure out how often we would succeed and how often we would fail in this example. Of the 17 possible samples of 1 that we might select, three samples would result in estimates of the population mean with an error greater than 3.0 cm, and 14 samples would result in estimates with an error of 3.0 cm or less. (The samples consisting of post hole No. 1, post hole No. 16, and post hole No. 17 would have means different from the known population mean by more than 3.0 cm.) Thus, 82.4% of the samples of 1 would provide us with estimates as accurate as we needed, but 17.6% of them would not.

If we selected samples of 1 over and over again, then, 82.4% of the time we would get a sample yielding an acceptably accurate estimate of the population mean, and 17.6% of the time we would get a sample yielding an unacceptably inaccurate estimate. (At least this is the case if each of these different samples is equally likely to occur, which, of course is true if the samples are randomly selected.) These percentages translate directly into *probabilities* for any single instance of drawing a sample of 1. That is, if 82.4% of the samples of 1 that we might draw would yield an acceptably accurate estimate of the population mean, then the *probability* of arriving at an acceptably accurate estimate in any single instance of drawing a sample of 1 would be 82.4% (or 0.824).

Stating the probability of occurrence of a single event in this manner means nothing more than stating the percentage of occurrence of that single event in a long sequence of repeated trials. We are accustomed to making such statements as, for example, when we say that the probability that a tossed coin will turn up heads is 50%. In saying this, we mean that when we toss a coin repeatedly, 50% of the time the result is heads. On a single toss the result will be either heads or tails, not half heads and half tails, but the probability of heads on a single toss is 50% because in repeated trials, 50% of the time the result will be heads and 50% of the time the result will be tails. This way of talking about probabilities is largely a matter of common sense and well established in common speech, but its importance to statistics is such that it merits explicit statement here.

In the example of drawing samples of 1 from a population of 17 post holes, then, we would achieve successful (that is, acceptably accurate) results 82.4% of the time. We would fail to attain the accuracy needed 17.6% of the time. If this success rate were not high enough, common sense tells us that we might do better with a larger sample.

ALL POSSIBLE SAMPLES OF A LARGER GIVEN SIZE

Suppose we selected samples of two post holes each from the population of 17 post holes. The range of possible results here is much larger. Our sample of 2 might consist of post hole No. 1 and post hole No. 1. (We are sampling here *with* replacement as discussed in Chapter 7.) Or our sample might consist of post hole No. 1 and post hole No. 2; or of post hole No. 1 and post hole No. 3; or of post hole No. 2 and post hole No. 3; and so on. In all there are 153 possible different samples of two post holes that could be selected from the population of 17 post holes (with replacement). If the samples were randomly selected, each of these 153 possible different samples of 2 would be equally likely to occur on any given drawing.

Of these 153 possible different samples of 2, some, of course, would give us estimates of the population mean with the level of accuracy we need (an error no greater than 3.0 cm) and some would not. It is not too difficult to determine which ones. The sample consisting of post hole No. 1 and post hole No. 1 would give a mean diameter of 10.4 cm, more than 3.0 cm in error. The next smallest possible sample mean would come from the sample consisting of post hole No. 1 and post hole No. 2. Here the mean diameter for the sample would be 10.55 cm. This is 2.98 cm less than the true population mean, so the error is acceptably small. There is, then, only one possible sample of 2 that would give an estimate of the population mean more than 3.0 cm too small.

At the other end of the scale sample means more than 3.0 cm higher than the population mean would be produced by all of the following samples of 2:

Post holes No. 17 and No. 17 ($\overline{X} = 20.30$ cm)
Post holes No. 17 and No. 16 ($\overline{X} = 19.35$ cm)
Post holes No. 17 and No. 15 ($\overline{X} = 18.35$ cm)
Post holes No. 17 and No. 14 ($\overline{X} = 17.65$ cm)
Post holes No. 17 and No. 13 ($\overline{X} = 17.30$ cm)
Post holes No. 17 and No. 12 ($\overline{X} = 17.15$ cm)
Post holes No. 17 and No. 11 ($\overline{X} = 17.00$ cm)
Post holes No. 17 and No. 10 ($\overline{X} = 16.75$ cm)
Post holes No. 17 and No. 9 ($\overline{X} = 16.60$ cm)
Post holes No. 16 and No. 16 ($\overline{X} = 18.40$ cm)
Post holes No. 16 and No. 15 ($\overline{X} = 17.40$ cm)
Post holes No. 16 and No. 14 ($\overline{X} = 16.70$ cm)

All the remaining possible samples of 2 that we might select would yield means no more than 3.0 cm different from the true population mean and would thus be acceptably accurate.

In sum, then, of the 153 possible different samples of 2 that we might select from the population of 17 post holes, 1 sample would yield an unacceptably low estimate of the population mean, 12 samples would yield unacceptably high estimates of the population mean, and 140 samples would yield acceptably accurate estimates of the population mean. Thus 140/153 (91.5%) of the time we would achieve successful (that is, acceptably accurate) results, and 8.5% of the time we would fail to achieve acceptable accuracy in estimating the population mean. Our probability of success on any given sample selection, then, is substantially greater with samples of 2 (acceptable accuracy 91.5% of the time) than it is with samples of 1 (acceptable accuracy 82.4% of the time). Samples of 2 that give unacceptably inaccurate results are more unusual than are samples of 1 that give unacceptably inaccurate results. Thus it is less likely that any particular random sample of 2 that we might select would give us unacceptably inaccurate results than was the case for samples of 1. The probability that any particular random sample of 2 yields unacceptably inaccurate results is 8.5% (or 0.085) in contrast to the probability of 17.6% (or 0.176) that any particular random sample of 1 would yield unacceptably inaccurate results. This is because such unrepresentative samples are more unusual among all possible samples of 2 than among all possible samples of 1.

If we extend the example to samples of 3, the same trend continues. There are 2,601 possible different samples of 3 that we might select from the population of 17 post holes. Of these, the following would yield estimates of the population mean more than 3.0 cm too low:

Post holes No. 1, No. 1, and No. 1 ($\overline{X} = 10.40$ cm)
Post holes No. 1, No. 1, and No. 2 ($\overline{X} = 10.50$ cm)

In addition, the following samples of 3 would yield estimates of the population mean more than 3.0 cm too high:

Post holes No. 17, No. 17, and No. 17 ($\overline{X} = 20.30$ cm)
Post holes No. 17, No. 17, and No. 16 ($\overline{X} = 19.67$ cm)
Post holes No. 17, No. 17, and No. 15 ($\overline{X} = 19.00$ cm)
Post holes No. 17, No. 17, and No. 14 ($\overline{X} = 18.53$ cm)
Post holes No. 17, No. 17, and No. 13 ($\overline{X} = 18.30$ cm)
Post holes No. 17, No. 17, and No. 12 ($\overline{X} = 18.20$ cm)
Post holes No. 17, No. 17, and No. 11 ($\overline{X} = 18.10$ cm)
Post holes No. 17, No. 17, and No. 10 ($\overline{X} = 17.93$ cm)
Post holes No. 17, No. 17, and No. 9 ($\overline{X} = 17.83$ cm)
Post holes No. 17, No. 17, and No. 8 ($\overline{X} = 17.73$ cm)
Post holes No. 17, No. 17, and No. 7 ($\overline{X} = 17.60$ cm)
Post holes No. 17, No. 17, and No. 6 ($\overline{X} = 17.43$ cm)
Post holes No. 17, No. 17, and No. 5 ($\overline{X} = 17.40$ cm)
Post holes No. 17, No. 17, and No. 4 ($\overline{X} = 17.37$ cm)
Post holes No. 17, No. 17, and No. 3 ($\overline{X} = 17.23$ cm)
Post holes No. 17, No. 17, and No. 2 ($\overline{X} = 17.10$ cm)
Post holes No. 17, No. 17, and No. 1 ($\overline{X} = 17.00$ cm)
Post holes No. 17, No. 16, and No. 16 ($\overline{X} = 19.03$ cm)
Post holes No. 17, No. 16, and No. 15 ($\overline{X} = 18.37$ cm)
Post holes No. 17, No. 16, and No. 14 ($\overline{X} = 17.90$ cm)
Post holes No. 17, No. 16, and No. 13 ($\overline{X} = 17.67$ cm)
Post holes No. 17, No. 16, and No. 12 ($\overline{X} = 17.57$ cm)
Post holes No. 17, No. 16, and No. 11 ($\overline{X} = 17.47$ cm)
Post holes No. 17, No. 16, and No. 10 ($\overline{X} = 17.30$ cm)

Post holes No. 17, No. 16, and No. 9 ($\overline{X} = 17.20$ cm)
Post holes No. 17, No. 16, and No. 8 ($\overline{X} = 17.10$ cm)
Post holes No. 17, No. 16, and No. 7 ($\overline{X} = 16.97$ cm)
Post holes No. 17, No. 16, and No. 6 ($\overline{X} = 16.80$ cm)
Post holes No. 17, No. 16, and No. 5 ($\overline{X} = 16.77$ cm)
Post holes No. 17, No. 16, and No. 4 ($\overline{X} = 16.73$ cm)
Post holes No. 17, No. 16, and No. 3 ($\overline{X} = 16.60$ cm)
Post holes No. 17, No. 15, and No. 15 ($\overline{X} = 17.70$ cm)
Post holes No. 17, No. 15, and No. 14 ($\overline{X} = 17.23$ cm)
Post holes No. 17, No. 15, and No. 13 ($\overline{X} = 17.00$ cm)
Post holes No. 17, No. 15, and No. 12 ($\overline{X} = 16.90$ cm)
Post holes No. 17, No. 15, and No. 11 ($\overline{X} = 16.80$ cm)
Post holes No. 17, No. 15, and No. 10 ($\overline{X} = 16.63$ cm)
Post holes No. 17, No. 14, and No. 14 ($\overline{X} = 16.76$ cm)
Post holes No. 16, No. 16, and No. 16 ($\overline{X} = 18.40$ cm)
Post holes No. 16, No. 16, and No. 15 ($\overline{X} = 17.73$ cm)
Post holes No. 16, No. 16, and No. 14 ($\overline{X} = 17.27$ cm)
Post holes No. 16, No. 16, and No. 13 ($\overline{X} = 17.03$ cm)
Post holes No. 16, No. 16, and No. 12 ($\overline{X} = 16.93$ cm)
Post holes No. 16, No. 16, and No. 11 ($\overline{X} = 16.83$ cm)
Post holes No. 16, No. 16, and No. 10 ($\overline{X} = 16.67$ cm)
Post holes No. 16, No. 16, and No. 9 ($\overline{X} = 16.57$ cm)
Post holes No. 16, No. 15, and No. 15 ($\overline{X} = 17.07$ cm)
Post holes No. 16, No. 15, and No. 14 ($\overline{X} = 16.60$ cm)

Thus 2 of the 2,601 possible samples of 3 would yield unacceptably low estimates and 48 would yield unacceptably high estimates. The acceptable accuracy rate would be 2,551/2,601; or 98.1%. The probability of selecting a random sample of 3 from this population of post holes that would yield an unacceptably inaccurate estimate of the population mean, then, is only 1.9% (or 0.019). This is because random samples of 3 with sample means so different from the mean of the population from which they were selected are fairly unusual (representing only 1.9% of the possible samples). It is thus very likely (not certain but very likely) that any particular sample of 3 that we might select from the population would represent the population with the accuracy we decided was needed in this example.

We could continue this example by considering the 44,217 possible different samples of 4 that could be selected, but the point should by now be clear. The larger the random sample is, the greater the chance that it represents the population from which it is selected with acceptable accuracy. Other things being equal, it is the size of the sample that governs its likely representativeness. Larger samples are more often representative of their parent populations than small samples. But, as has been emphasized above, large samples provide *no guarantee* of representativeness. The most unrepresentative sample of 3 in this example consists of post hole No. 17 selected three times. This sample is just exactly as unrepresentative as the most unrepresentative sample of 1 (consisting of post hole No. 17). But such unrepresentative samples occur far less frequently among larger samples than among smaller samples.

The number of errors of more than 3.0 cm in estimating the mean in the population of 17 post hole diameters also depends on the spread of the population. If there are many post holes much larger or much smaller than the mean, then the number of samples producing unacceptably inaccurate results increases. If this does not initially make sense to you, go back to the example population given in Table 8.1 and change post holes 1, 2, and 3 to 9.0 cm, 9.4 cm, and 9.8 cm, respectively. Start counting up how many samples of 1, 2, and 3 there would be with means more than 3.0 cm different from 13.53 cm. The bigger the spread in the population, the more samples there will be whose means are not acceptably close to the true population mean (for any given definition of "acceptably close").

The chance of making badly erroneous inferences about populations on the basis of samples, then, is less with larger samples, although a small risk of serious error remains even with large samples. The chance of making badly erroneous inferences about populations on the basis of samples is also less when the population is homogeneous (a batch with a small spread) and greater when the population is highly variable (a batch with a larger spread). In this specific example, in which we know exactly what the population is like, and we established (even if arbitrarily) what "acceptable" accuracy was, we could easily figure the percentages of samples that would yield acceptable and unacceptable results. What we need now is a means of generalizing the observations that we made in this specific example.

THE "SPECIAL BATCH"

The key to general application of the specific observations we made in the example above lies in a very special batch of numbers. *This "special batch" consists of the means of all the possible different samples of a given size that could be drawn from a given population.* Let's consider this in terms of the previous example.

For a sample size of 1 (that is, for $n = 1$), there are 17 different random samples that could be selected from our example population of 17 post holes. Each of the 17 samples has its own sample mean ($\overline{X}$). The special batch would consist of these 17 sample means. We found earlier that 17.6% of these 17 sample means were more than 3.0 cm different from the real population mean, and they were therefore classified as unacceptably unrepresentative samples. Unacceptably unrepresentative samples of 1 were thus a bit unusual, making up only 17.6% of the special batch, but we would not call them extremely rare. The clear majority of the samples of 1 that we could select from this population would represent it with sufficient accuracy for our present purposes, but an uncomfortably large proportion of the samples of 1 we might select would be unacceptably inaccurate.

For $n = 2$, there are 153 different random samples that could be selected from our example population of 17 post holes. Each of these 153 samples has its own sample mean ($\overline{X}$). The special batch would consist of these 153 sample means. Samples so unrepresentative that their means differed by more than 3.0 cm from the population mean were more unusual in terms of this special batch, making up only 8.5% of the possible samples of 2 that could be selected from this population.

For $n = 3$, there are 2,601 different random samples that could be selected from our example population of 17 post holes. Each of these 2,601 samples has its own sample mean ($\overline{X}$). The special batch would consist of these 2,601 sample means. Unacceptably unrepresentative samples were even more unusual among samples of 3, making up only 1.9% of the special batch.

And we could go on. For a given population and for any given sample size, there is a special batch consisting of the means of all the different samples of that size that could be selected randomly from that population. This special batch, then, consists of all the possible results we could obtain in estimating the given population's mean on the basis of a sample of the given size. And this special batch is the key to determining just how unusual it would be to draw an unacceptably unrepresentative sample of a certain size from the given population. The unusualness of an unacceptably unrepresentative sample (in terms of the special batch) enables us to specify the probability that any specific sample of a given size that we might randomly select from a given population will be unrepresentative.

THE STANDARD ERROR

We have just been using the notion of unusualness in very much the same way we used it in Chapter 4 – unusualness of a number in terms of the batch of numbers to which it belongs. Since the numbers we have been discussing are the means of samples of particular sizes, the comparison batch has been the batch consisting of all the means of samples of a given size from a given population, that is, the special batch. In Chapter 4 we talked about more general tools for evaluating the unusualness of a number in terms of its batch, tools based on numerical indexes of the level and spread of the batch. We could use just such tools in this effort to discuss unusualness of sample results in terms of the special batch. In order to do so we would need to know the level and spread of the special batch. We could, of course, find out the level and spread of the special batch by selecting all possible samples of a given size and working directly with the batch, but this is obviously preposterous. It would be considerably more work than just studying whatever we wanted to study in the whole population, and so sampling would offer no advantage. It turns out that there are much easier ways to find out about the special batch.

It can be shown mathematically that *the mean of the special batch is the same as the mean of the population from which the samples were drawn*. This, of course, is quite apparent in the case of samples of 1. The special batch, for samples of 1, is exactly the same batch of numbers as the population, since each sample is the same as one number in the population. The mean of the population, then, has to be the same as the mean of the special batch. It turns out that this is true even when $n > 1$ (that is, even when the sample size is greater than 1).

If we can say that the mean of this special batch is the same as the mean of the population from which the samples are drawn, then we can say that the mean of the means of all the possible samples of a given size that can be drawn from a given population is the same as the mean of that population. These two statements are synonymous because the special batch *is* the means of all the possible samples of a given size that can be drawn from a given population.

You can actually think this through fairly easily for yourself if you want to, without need of formal mathematical proofs. If we select all possible samples of any given size, each number in the population occurs an equal number of times in all the samples taken together (however many times that may be – it depends on the sample size). The mean of all the sample means is also the mean of all the numbers in all the samples, taken as one immense undivided batch. Since all numbers in the population occur the same number of times in all the samples taken together, this immense batch is simply the original population reduplicated many times over, and its mean will be the same as the mean of the original population. Each number has simply been added in many times, but then the total has been divided by a much larger number, reflecting precisely how many times each number has been added in.

It can also be shown mathematically that *the standard deviation of the special batch is the standard deviation of the given population divided by the square root of the number of elements in the sample.* The truth of this is, once again, obvious when the sample size is 1. The standard deviation of the special batch is the standard deviation of the population divided by the square root of 1 (the sample size). Since the square root of 1 is 1, the standard deviation of the special batch is the same as the standard deviation of the population when the sample size is 1. This is not surprising, since the special batch is the same as the population when the sample size is 1. This same relationship, however, between the standard deviation of the special batch, the sample size, and the standard deviation in the population holds true for any given sample size.

The standard deviation of the special batch is such an important number that it has its own special name. It is the standard error. *The standard error, then, is the standard deviation of the batch consisting of the means of all the different samples of a given size that could be selected from a given population.* The equation for standard error is

$$SE = \frac{\sigma}{\sqrt{n}}$$

where SE = standard error, and σ = standard deviation of the population, and n = number of elements in the sample.

We are now in position to specify a numerical index of level and a numerical index of spread for the special batch so as to discuss the unusualness of particular samples in a general and efficient way. The numerical indexes we have specified, however, are two that we have seen behave very badly in previous chapters. Neither mean nor standard deviation is at all resistant to the effect of outliers or asymmetry. Here we are in luck, however, because, for samples of relatively large size, it can also be shown mathematically that the shape of the special batch is normal. Since normal shapes are single peaked and symmetrical, we know that the mean and standard deviation will be useful numerical indexes of level and spread, and we do not have to worry about the fact that they are not resistant. Relatively large sample size, in this instance, can be taken to mean more than about 30. This characteristic of the special batch (having a normal shape for relatively large sample size) is also of pivotal importance. It is called the *central limit theorem*.

To summarize, in this section we have conceived of a special batch of numbers that consists of the means of all the different samples of a given size that could be drawn from a given population. This special batch is known in more formal statistical terminology as the *sampling distribution of the mean*, but we will continue to refer to it here simply as the special batch. Three properties of the special batch have been noted. First, the mean of the special batch is the same as the mean of the population from which the samples are selected. Second, the standard deviation of the special batch, known as the standard error of the sample, is $\sigma/\sqrt{n}$. And third, the shape of the special batch is normal as long as the sample size is over about 30.

These three properties of the special batch give us rather complete information about its characteristics. Without having to actually select and manipulate all possible samples of a given size, we can determine the level (mean), spread (standard deviation), and shape (single peaked, symmetrical, normal) of the special batch. In the next chapter we will put the special batch and its characteristics to general use in assessing the unusualness of particular samples.

Chapter 9
Confidence and Population Means

The major difficulty in putting the properties of the special batch discussed in Chapter 8 to use is that we had to know a good deal about the population from which the sample was drawn in order to specify the characteristics of the special batch. We knew that the mean of the special batch was the same as the mean of the population and that the standard deviation of the special batch (that is, the standard error of the sample) was the standard deviation of the population divided by the square root of the number in the sample. In real life, however, we do not know either the mean or the standard deviation of the population from which our sample is drawn. Indeed those are precisely the things we are trying to estimate on the basis of a sample. Thus we must find a way to use the special batch without first knowing these characteristics of the entire population.

In this chapter we will extend the notion of unusualness of a sample to apply to the more realistic situation in which, instead of having one population and all the possible samples from it, we have one sample and consider the possible populations it might have come from. We will start by asking the question, "How unusual would it be for the sample we actually have to come from a population with a particular mean?" And we will proceed to ask that question about a number of different possible parent populations for our sample.

R.D. Drennan, *Statistics for Archaeologists*, Interdisciplinary Contributions
to Archaeology, DOI 10.1007/978-1-4419-0413-3_9,
© Springer Science+Business Media, LLC 2004, 2009

GETTING STARTED WITH A RANDOM SAMPLE

Let's suppose that we have a random sample of 100 projectile points drawn from a much larger population of projectile points, whose mean length we wish to know. This random sample of 100 projectile points has a mean length of 3.35 cm and a standard deviation of 0.50 cm. Such a situation may occur in real life when, for example, we have surveyed a region intensively and made systematic surface collections at all the sites encountered. To keep the logic simpler, let's suppose that study of these collections revealed occupation of the region during only a single prehistoric period. We decide to take all the projectile points recovered in these collections (100 points altogether) as a random sample from the population consisting of all the projectile points made by the prehistoric inhabitants of the region during the single period during which the region was occupied.

Our sample is not technically a random sample, but we might decide to treat it as one, at least for estimating the mean projectile point length in the population. In order to make this decision we would need to consider the collecting procedures used in the field as well as the processes by which projectile points are brought to the surfaces of sites and become available for collection. These latter processes include the full range of things that happen to projectile points from the time they are discarded to the time they are found. If, in considering all these processes, we can find no reason to believe that projectile points of different lengths will be affected in substantially different ways (or at least that whatever such effects may be, they apply equally to this sample and to other samples with which we wish to compare this sample), then we would proceed to treat this sample as a random sample with respect to projectile point length. The legitimacy of any conclusions we make about projectile point length in the population, of course, is dependent on this decision. We must recognize the possibility in using these conclusions that, at some time in the future, they might be invalidated if we were to discover that the sample had been biased with respect to projectile point length in some way we had not thought of.

This procedure may seem risky, but, as discussed in Chapter 7, the only alternative is simply not to make conclusions about projectile point length in the larger population. Whatever statements we make about, say, Late Woodland projectile points in general are based on precisely such logic, whether those statements are statistical in nature or purely subjective impressions. Archaeologists have always made such general statements about large and vaguely defined populations on the basis of samples not randomly selected. And such statements, even when statistics have been in no way involved, are based on treating the sample at hand as if it were not biased even when we cannot show conclusively that bias is absent. This approach is no more risky when it serves as the foundation for statistical statements than when it serves as the foundation for subjective impressions. Indeed it is less risky. This is because the statistical techniques we are about to apply only assume that the sample is unbiased; they do not assume that it accurately represents the population from which it came, only that it is not systematically biased. Subjective generalizations assume not only that the sample upon which they are based is unbiased, but also that the sample provides completely accurate representation – a stronger assumption, and one much more difficult to justify.

Archaeologists are not the only scientists in this situation. We are all comfortable using such figures as the mean heights of adult males and adult females in the United States. We seldom even think about where such figures come from. Clearly they do not involve measuring the heights of all adult males and all adult females in the country. The figures are based on a much smaller sample. Even that is not technically a random sample of all adult males and all adult females in the country. It was a sample from a much smaller subpopulation that was simply *taken to accurately represent the larger population.* No one ever actually assigned numbers to every adult male and every adult female in the United States, randomly selected a sample, and set out to measure every individual in the sample. Much smaller and more accessible populations were taken to accurately represent the nation's population at large after careful consideration and elimination of the ways in which such populations might be biased samples.

In exactly the same way, archaeologists do not need to be able to number sequentially and randomly select a sample from all the projectile points made in a particular period in a particular region in order to characterize this large and vaguely defined population. Archaeologists can (and must) argue that the projectile points lying on the surface at a given moment are an unbiased subgroup of that larger population (with respect to certain characteristics at least) and that the 100 projectile points recovered on survey are an unbiased sample from that subgroup. This is the way sampling of such large and vaguely defined populations is customarily done in many disciplines. The conclusions produced are reliable only to the extent that the assumption that the sample is unbiased can be justified. If this is in doubt, then this doubt remains as a doubt about the validity of the conclusion reached.

As long as we have digressed from the topic at hand to such a lengthy discussion of the real-life implications of the assumptions of sampling, we might as well specify one terminological point as well. Large and vaguely defined populations like the one we are dealing with here are referred to in statistics as *infinite populations.* This does not mean that they are truly infinite, just that they are very large and not precisely defined. (We will see as we continue to discuss the notion of infinite populations that infinity is a much smaller thing to a statistician than to an astronomer.)

WHAT POPULATIONS MIGHT THE SAMPLE HAVE COME FROM?

Once we've satisfied ourselves that we are willing to treat the sample we have as if it were a random sample (at least for purposes of argument), we can begin to consider what kind of population the sample might have come from. Recall that our sample of 100 projectile points had a mean length of 3.35 cm and a standard deviation of 0.50 cm. For large populations and large samples, the sample mean is the same as the population mean more often than it is any other one figure. Similarly, the sample standard deviation is the same as the population standard deviation more often than it is any other one figure. Thus our best estimate is that the population of projectile points from which this sample was selected has a mean length of 3.35 cm and a standard deviation of 0.50 cm.

We know, however, that samples do not always have exactly the same mean as their parent populations, so we wonder just how much confidence we should have in this estimate. Put another way, just how likely is it that this estimate is incorrect? Put more fully, just how likely is it that this estimate is incorrect by enough to matter? The addition of that last phrase is an important practical matter of precision. We almost certainly do not need to worry about the possibility that the real population mean might be 3.350000001 cm as opposed to 3.350000000 cm. This difference of 0.000000001 cm is clearly not enough to matter. It is almost certainly well beyond the capability of our measuring instruments to even detect such a difference. But the point is that we do not seek infinite precision even if it were possible – it wouldn't matter. Being incorrect by enough to matter is what we have to worry about. Probably 0.01 cm or even 0.1 cm is not enough to worry about. Maybe even 0.4 cm or 0.5 cm is not a large enough error in estimating the population mean to worry us seriously.

The question of necessary precision is not one of applying statistical rules of precision. Rather it is a substantive question involved with why we want to know what the mean length of projectile points in this population is. For statistical purposes, then, we take whatever decision is made about necessary precision as a given because that decision is based on substantive concerns outside the realm of statistics. For example, our reason for wanting to know the mean length of projectile points in our region may be to compare this length with the mean length for another region in an effort to determine something about differences in hunting practices. In this case, a difference of 0.1 cm would likely not be taken as meaningful in that it would seem too small to be reflecting a meaningful difference in hunting practices. A difference of 0.5 cm might, on the other hand, be meaningful, if a substantive case could be made for what, specifically, it would indicate.

Turning back to the sample that we have, we have already guessed that it most likely came from a population with a mean length of 3.35 cm (the same as the sample mean). But we know that there is no guarantee that it came from such a population. Our sample might have come from a population with a mean length greater or less than 3.35 cm, possibly even from a population with a mean length much greater or less than 3.35 cm. We can begin to think about how likely this is by considering various specific populations from which our sample might have come. For each specific population we imagine that our sample might have come from, we will need to think of the special batch consisting of the means of all possible samples of 100 from that population.

For starters, let's imagine our sample might have come from a population with a mean length of 3.25 cm. How unusual would it be to get a sample like ours (that is, with a mean of 3.35 cm and a standard deviation of 0.50 cm) from a population with a mean of 3.25 cm? What would the special batch consisting of the means of all

possible samples of 100 from a population with a mean of 3.25 cm look like? We
know that the mean of this special batch would be the same as the population mean,
that is, 3.25 cm. We know that the shape of this special batch would be approximately normal because of the central limit theorem and because 100 is a fairly large
sample. We only lack knowledge of the spread of the special batch, but we know
that the spread of the special batch is given by the equation

$$SE = \frac{\sigma}{\sqrt{n}}$$

Since we have no better recourse, we will continue to use the standard deviation
of the sample (0.50 cm) as our best estimate of the standard deviation in the parent
population. Thus

$$SE = \frac{0.50\,cm}{\sqrt{100}} = \frac{0.50\,cm}{10} = 0.05\,cm$$

Figure 9.1 illustrates the special batch consisting of the means of all the possible
samples of 100 that could be drawn from a population with a mean of 3.25 cm and
a standard deviation of 0.50 cm. This is simply a histogram, like those discussed
in Chapter 1. Clearly, samples with means close to 3.25 cm are much more common than are samples with means far from 3.25 cm. Figure 9.2 illustrates this same
special batch in a more common and useful manner. Instead of a histogram with specific intervals represented by vertical bars, the heights of the bars are represented by
a smooth curve joining the center points of the tops of the bars. This allows us to use
the horizontal scale as the truly continuous measurement scale that it is instead of
breaking it up into awkward intervals. The height of the curve above any point along
the horizontal scale, then, represents the frequency with which samples with a particular mean occur, in just the same way that the height of a bar on the corresponding

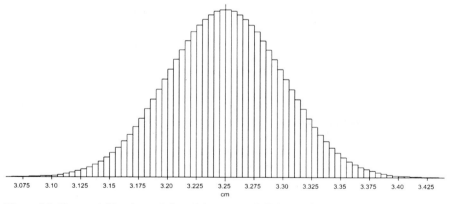

Figure 9.1. The special batch consisting of the means of all the possible samples of 100 that could
be selected from a population with a mean of 3.25 cm and a standard deviation of 0.50 cm.

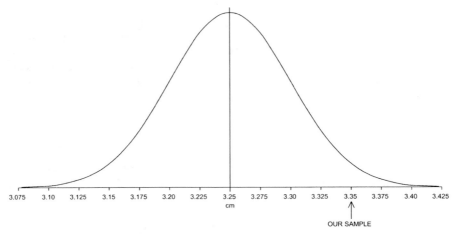

| 3.075 | 3.10 | 3.125 | 3.15 | 3.175 | 3.20 | 3.225 | 3.25 | 3.275 | 3.30 | 3.325 | 3.35 | 3.375 | 3.40 | 3.425 |

cm

OUR SAMPLE

Figure 9.2. The special batch for samples of 100 from a population with a mean of 3.25 cm and a standard deviation of 0.50 cm.

histogram represents the frequency of occurrence of samples with a mean falling in a particular interval. Representing the shape of a batch with a normal distribution in this way is so common in statistics that the entire concept is often referred to in a kind of shorthand as the "normal curve."

For a given mean (in this case 3.25 cm) and a given standard deviation (in this case the standard error, which is the standard deviation of the special batch, or 0.05 cm) there is one and only one specific normal distribution, and Fig. 9.2 is it. Figure 9.2 is thus a picture of the special batch consisting of the means of all possible samples of 100 that can be selected from a population with a mean of 3.25 cm and a standard deviation of 0.50 cm. We can use this picture to place our sample, with a mean of 3.35 cm, in context with all other possible samples. The position of our sample in this distribution is indicated in Fig. 9.2. At the point corresponding to our sample, the normal curve is fairly low, indicating that samples with a mean of 3.35 cm do occur among the possible samples of 100 from a population with a mean of 3.25 cm, but they do not occur very frequently – not nearly as frequently, for example, as samples with means closer to 3.25 cm. Our sample is fairly unusual, then, in the context of all the possible samples from a population with a mean of 3.25. It is therefore possible, but not very likely, that our sample came from a population with a mean of 3.25 cm.

We can do the same thing for other populations from which our sample might possibly have come. For instance, how likely is it that our sample came from a population with a mean length of 3.20 cm? Figure 9.3 illustrates the special batch consisting of the means of all possible samples of 100 that could be selected from a population with a mean of 3.20 cm and a standard deviation of 0.50 cm. The level of the normal curve at the point corresponding to our sample in Fig. 9.3 is extremely low. Thus our sample, with its mean of 3.35 cm, would be extremely unusual among

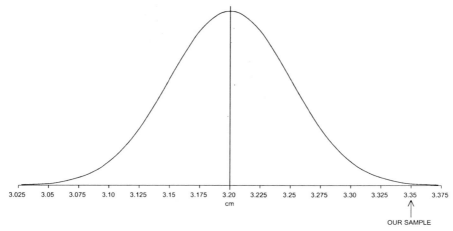

Figure 9.3. The special batch for samples of 100 from a population with a mean of 3.20 cm and a standard deviation of 0.50 cm.

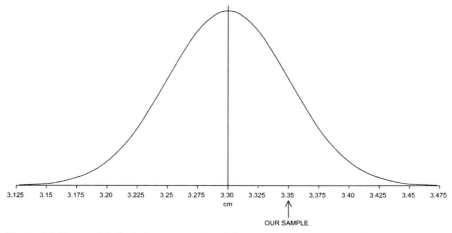

Figure 9.4. The special batch for samples of 100 from a population with a mean of 3.30 cm and a standard deviation of 0.50 cm.

samples of 100 selected from a population with a mean of 3.20 cm. It is therefore very unlikely (although not entirely impossible) that our sample came from such a population.

How likely is it that our sample came from a population with a mean of 3.30 cm? Figure 9.4 illustrates the special batch consisting of the means of all possible samples of 100 that could be selected from a population with a mean of 3.30 cm and a standard deviation of 0.50 cm. The level of the normal curve at the point corresponding to our sample in Fig. 9.4 is fairly high. Thus there are a good many

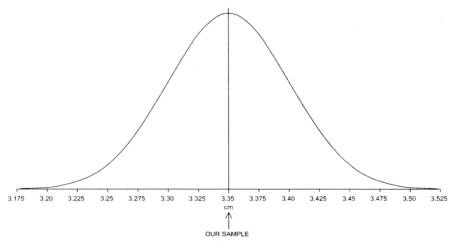

3.175   3.20   3.225   3.25   3.275   3.30   3.325   3.35   3.375   3.40   3.425   3.45   3.475   3.50   3.525

cm

↑

OUR SAMPLE

Figure 9.5. The special batch for samples of 100 from a population with a mean of 3.35 cm and a standard deviation of 0.50 cm.

samples like ours among those possible to select from a population with a mean of 3.30 cm. Therefore it is relatively likely that our sample could have come from such a population.

Finally, Fig. 9.5 illustrates the special batch corresponding to the population with a mean of 3.35 cm – the population that is a more likely parent population for our sample than any other single population. We could imagine continuing to try out many more possible parent populations in this way and constructing a new curve from the results of these trials. This new curve would indicate how likely it was that each of the possible parent populations was indeed the population from which our sample was drawn. It turns out that if we carried out this procedure, the curve we would construct would have exactly the same parameters as the curve illustrated in Fig. 9.5. In effect what we have done is to turn the logic of the curve in Fig. 9.5 inside out to produce the curve in Fig. 9.6.

Figure 9.5, again, represents the special batch composed of the means of all the possible samples of 100 that could be selected from a population with a mean of 3.35 cm and a standard deviation of 0.50 cm. It thus represents the unusualness of the various samples that could be selected from this population and therefore the probability of selecting any one of them from this population. Figure 9.6, on the other hand, represents the means of the possible populations that a sample of 100 with a mean of 3.35 cm and a standard deviation of 0.50 cm might have been drawn from and therefore the probability that this sample was selected from any particular one of them. This batch, represented in Fig. 9.6, has exactly the same level, spread, and shape as the special batch that we have been discussing. That is, just like the familiar special batch, this second batch has a mean that is the same as the sample mean; it has a standard deviation that is $\sigma/\sqrt{n}$ or the standard error; and its shape is normal.

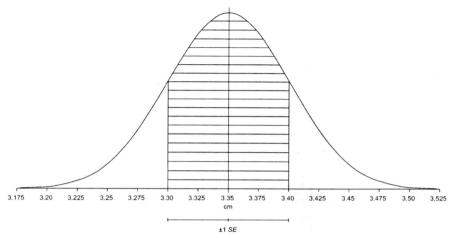

3.175   3.20   3.225   3.25   3.275   3.30   3.325   3.35   3.375   3.40   3.425   3.45   3.475   3.50   3.525
cm

±1 SE

Figure 9.6. The batch consisting of the means of the populations from which a sample of 100 with a mean of 3.35 cm and a standard deviation of 0.50 cm might have come. The majority of the means lie within 1 standard error of the sample mean, but a substantial number of means are larger or smaller than this.

CONFIDENCE VERSUS PRECISION

We can look at Fig. 9.6 and quickly say that a good many of the populations that our sample might have come from have means between 3.30 cm and 3.40 cm. (These are the populations that fall within 1 standard error of the mean of our sample.) According to the shape of the special batch, however, a good many of the possible populations have means outside that range. Thus we are only moderately confident that the population our sample came from has a mean between 3.30 cm and 3.40 cm. We say this because populations with means less than 3.30 cm or greater than 3.40 cm are relatively numerous among the possible populations. It would not strain credulity at all to imagine selecting a sample with a mean of 3.35 cm and a standard deviation of 0.50 cm from a population with a mean less than 3.30 cm or greater than 3.40 cm. Figure 9.6 shows us that such a thing would happen with some frequency. Thus our sample probably came from a population with a mean between 3.30 cm and 3.40 cm, but there is a very real chance that it might not have. It means the same thing to say, "The probability is moderate that our sample came from a population with a mean of 3.35 cm ± 0.05 cm."

Suppose we are not satisfied with the lack of confidence we have in the statement that the population our sample came from probably has a mean between 3.30 cm and 3.40 cm. We can speak more confidently, but only by reducing the level of precision of our statement. We could say that the population our sample came from has a mean between 3.25 cm and 3.45 cm, and be somewhat more confident that our statement is true. This statement is illustrated by Fig. 9.7, where the clear majority of the possible populations have means that fall between 3.25 cm and 3.45 cm. It seems

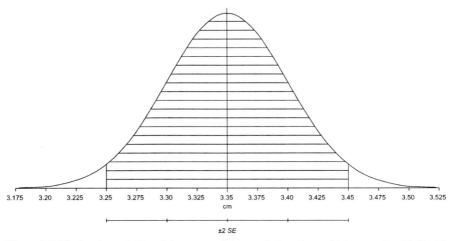

Figure 9.7. The batch consisting of the means of the populations from which a sample of 100 with a mean of 3.35 cm and a standard deviation of 0.50 cm might have come. The vast majority of the means lie within 2 standard errors of the sample mean.

quite likely that our sample comes from a population with a mean somewhere in this range. Relatively few of the possible populations fall outside the range. Thus it would be fairly unusual to select a sample like ours (with a mean of 3.35 cm and a standard deviation of 0.50 cm) from a population with a mean less than 3.25 cm or greater than 3.45 cm. The probability that our sample came from a population with a mean less than 3.25 cm or greater than 3.45 cm is low. Correspondingly, the probability that our sample came from a population with a mean between 3.25 cm and 3.45 cm is high. Thus we might say something like, "There is a high probability that our sample came from a population with a mean of 3.35 cm ± 0.10 cm." This statement indicates greater confidence than the statement at the end of the previous paragraph, but it is a less precise statement.

The twin notions of confidence and precision are familiar to us in common colloquial speech, although we usually don't think of them directly. If I intend to make quite sure I will arrive for an appointment at a precise time, I might say, "I will be there at 4 o'clock." Customs of punctuality vary, but I am not likely to say that unless I feel quite confident that I will arrive within about 5 minutes of 4 o'clock. If my arrival depends on how heavy traffic is en route, I am more likely to say, "I will be there about 4 o'clock," a less precise statement, indicating that I might be 10 or 15 minutes early or late. If I envision still more imponderable interference with my schedule, I might say, "I will be there sometime around 4 o'clock," indicating still less precision, perhaps between 3:30 and 4:30.

I could communicate similar messages by varying the confidence implied in my statements. Instead of saying "I will be there about 4 o'clock," I could say, "I will probably be there at 4 o'clock." The former statement encourages the listener to think of a period of 20 minutes or so during which my arrival can be expected. The latter statement instead encourages the listener to imagine the precise moment of 4 o'clock, but not to have too much confidence that I will be present then. The two statements convey very similar messages, but I might use them in different contexts. If I am going to a meeting with a colleague, which will begin when I arrive, I would say "I will be there about 4 o'clock," thinking of the range of time during which the meeting can be expected to begin. If, on the other hand, I am going to a lecture scheduled to start at 4 o'clock whether I am there or not, I would say, "I will probably be there at 4 o'clock," imagining how likely it is that I will be present at the precise time the lecture can be expected to begin. It is usually a trade-off between speaking with precision and speaking with confidence. Other things being equal, the more precision we speak with, the lower our confidence; and the more confidence we speak with, the less precise our statements. Only in unusual circumstances am I able to say, "I *will* be there at 4 o'clock sharp," emphasizing that I am speaking with both high confidence ("I *will*") and high precision ("4 o'clock sharp"). At the other end of the scale is both low confidence and low precision: "I'll see if I can be there sometime around 4 o'clock."

The statistical statements we are making about the kind of population our sample came from work in exactly the same way. We can either indicate very high confidence that the population has a mean in a somewhat imprecise range of values or indicate a population mean with greater precision but lower confidence that we're correct. Figure 9.8 continues the progression begun in Figs. 9.6 and 9.7. It illustrates a still less precise statement, but one that can be made with great confidence. Almost all the possible populations that a sample like ours (of 100 elements with a mean of 3.35 cm and a standard deviation of 0.50 cm) could come from have means that fall in the range between 3.20 cm and 3.50 cm. Very few of the possible populations have means less than 3.20 cm or greater than 3.50 cm. It would be quite unusual to

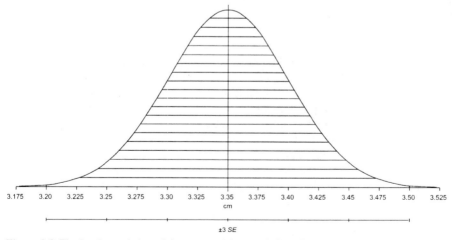

Figure 9.8. The batch consisting of the means of the populations from which a sample of 100 with a mean of 3.35 cm and a standard deviation of 0.50 cm might have come. Only a few means are more than 3 standard errors from the sample mean.

select a sample of 100 with a mean of 3.35 cm and a standard deviation of 0.50 cm from a population with a mean less than 3.20 cm or greater than 3.50 cm. Thus it is very unlikely that our sample came from a population with a mean less than 3.20 cm or greater than 3.50 cm. It is very likely that our sample came from a population with a mean between 3.20 cm and 3.50 cm. We could say, "The probability is very high that the population our sample came from has a mean of 3.35 cm ± .15 cm."

PUTTING A FINER POINT ON PROBABILITIES – STUDENT'S *t*

The notions of approximate probabilities we have been using thus far can be extended to much more precise and useful ways of assessing probabilities on the basis of how unusual a particular result would be in the context of all the possible results. We have used the approximate height of the normal curve (and thus the shaded areas enclosed by it in Figs. 9.6, 9.7, and 9.8) to judge roughly how unusual (and thus how improbable) it would be for our sample to have been selected from populations with means falling in different ranges. These ranges of possible means are called *error ranges* or *confidence intervals*. They are most often expressed as a "±" quantity following the mean. Figure 9.6 illustrates an error range of ±1 standard error; Fig. 9.7 illustrates an error range of ±2 standard errors; and Fig. 9.8 illustrates an error range of ±3 standard errors. We concluded earlier that we are very confident that the mean of the population our sample came from lies within the ±3 standard error range (Fig. 9.8); we are fairly confident that the mean of the population our sample came from lies within the ±2 standard error range (Fig. 9.7); and we have only modest confidence that the mean of the population our sample came from lies within the ±1 standard error range (Fig. 9.6).

The exact *levels of confidence* we have in these three statements of differing precision can be found by calculating the exact areas "under the normal curve" in Figs. 9.6, 9.7, and 9.8. Student's *t* distribution provides us with these exact areas. The key to use of Student's *t* is in numerical indexes of level and spread (in this case the mean and standard deviation) used to measure the unusualness of a particular number in a batch. The relevant batch is the special batch, whose mean is the same as the mean of our sample and whose standard deviation is the standard error of our sample. (Be sure not to confuse the standard deviation of the sample or of the population with the standard deviation of the special batch. The standard deviation of the special batch is the standard error of the sample.) Student's *t*, then, provides a detailed description of the shape of the special batch for us to use.

Figure 9.7, for example, illustrates an error range (3.35 cm ± 0.10 cm) consisting of 2 standard errors. We already recognized that it is very likely that the population our sample came from has a mean that falls within this error range. Table 9.1 allows us to say just what we mean by "very likely" in the following manner. First we must determine the row of the table to use, based on the size of the sample. The left-hand column indicates the *degrees of freedom*, which are equivalent to one less

Table 9.1. Student's *t* Distribution

CONFIDENCE (%)   (NO. OF STANDARD ERRORS)   SIGNIFICANCE (%)   (NO. OF STANDARD ERRORS)

Confidence	50%	80%	90%	95%	98%	99%	99.5%	99.8%	99.9%
	.5	.8	.9	.95	.98	.99	.995	.998	.999
Significance	50%	20%	10%	5%	2%	1%	.5%	.2%	.1%
	.5	.2	.1	.05	.02	.01	.005	.002	.001

Degrees of freedom

1	1.000	3.078	6.314	12.706	31.821	63.637	127.32	318.31	636.62
2	.816	1.886	2.920	4.303	6.965	9.925	14.089	22.326	31.598
3	.765	1.638	2.353	3.182	4.541	5.841	7.453	1.213	12.924
4	.741	1.533	2.132	2.776	3.747	4.604	5.598	7.173	8.610
5	.727	1.476	2.015	2.571	3.365	4.032	4.773	5.893	6.869
6	.718	1.440	1.943	2.447	3.143	3.707	4.317	5.208	5.959
7	.711	1.415	1.895	2.365	2.998	3.499	4.020	4.785	5.408
8	.706	1.397	1.860	2.306	2.896	3.355	3.833	4.501	5.041
9	.703	1.383	1.833	2.262	2.821	3.250	3.690	4.297	4.781
10	.700	1.372	1.812	2.228	2.764	3.169	3.581	4.144	4.537
11	.697	1.363	1.796	2.201	2.718	3.106	3.497	4.025	4.437
12	.695	1.356	1.782	2.179	2.681	3.055	3.428	3.930	4.318
13	.694	1.350	1.771	2.160	2.650	3.012	3.372	3.852	4.221
14	.692	1.345	1.761	2.145	2.624	2.977	3.326	3.787	4.140
15	.691	1.341	1.753	2.131	2.602	2.947	3.286	3.733	4.073
16	.690	1.337	1.746	2.120	2.583	2.921	3.252	3.686	4.015
17	.689	1.333	1.740	2.110	2.567	2.898	3.222	3.646	3.965
18	.688	1.330	1.734	2.101	2.552	2.878	3.197	3.610	3.922
19	.688	1.328	1.729	2.093	2.539	2.861	3.174	3.579	3.883
20	.687	1.325	1.725	2.086	2.528	2.845	3.153	3.552	3.850
21	.686	1.323	1.721	2.080	2.518	2.831	3.135	3.527	3.819
22	.686	1.321	1.717	2.074	2.508	2.819	3.119	3.505	3.792
23	.685	1.319	1.714	2.069	2.500	2.807	3.104	3.485	3.767
24	.685	1.318	1.711	2.064	2.492	2.797	3.091	3.467	3.745
25	.684	1.316	1.708	2.060	2.485	2.787	3.078	3.450	3.725
30	.683	1.310	1.697	2.042	2.457	2.750	3.030	3.385	3.646
40	.681	1.303	1.684	2.021	2.423	2.704	2.971	3.307	3.551
60	.679	1.296	1.671	2.000	2.390	2.660	2.915	3.232	3.460
120	.677	1.289	1.658	1.980	2.358	2.617	2.860	3.160	3.373
∞	.674	1.282	1.645	1.960	2.326	2.576	2.807	3.090	3.291

(Adapted from Table 3 in *Introduction to Contemporary Statistical Methods* by Lambert H. Koopmans (Boston, MA: Duxbury Press, 1987) than the number of elements in the sample ($n - 1$). For the moment, we will just take this notion of degrees of freedom (often abbreviated $d.f.$) on faith. For our sample, $n - 1 = 99$. There is no row corresponding exactly to 99 degrees of freedom, so we will use the row for 120 $d.f.$, which comes closest. We are looking for the exact level of confidence associated with an error range of 2 standard errors, so we read across that row looking for 2. In the fourth column we find 1.98 (which we'll take as close enough to 2 for the moment).

The fourth column is headed 95% confidence. This means that 95% of the possible populations (represented by the shaded area "under the normal curve" in Fig. 9.7) that our sample could come from lie within 1.98 standard errors of the mean of our sample. Thus, when we say that it is "very likely" that our sample came from a population with a mean of 3.35 cm ± 0.10 cm, what we mean more precisely is that there is about a 95% probability that our sample came from such a population. We are 95% confident that our sample came from a population with a mean of 3.35 cm ± 0.10 cm. We are *not certain* that our sample came from a population with a mean of 3.35 cm ± 0.10 cm, but the probability that this is the case is 95%.

Since the probability that our sample came from a population with a mean between 3.25 cm and 3.45 cm is 95%, the probability that it came from a population with a mean less than 3.25 cm or greater than 3.45 cm is 5%. (This has to be true since the probability that it came from one or the other of these groups is 100%.) Since a normal shape is symmetrical, this 5% is evenly distributed in both "tails" of the distribution. There is a 2.5% probability that our sample came from a population with a mean less than 3.25 cm and a 2.5% probability that our sample came from a population with a mean greater than 3.45 cm. When we provide an error range of about 2 standard errors, then, as we have done here, we are speaking at a *95% confidence level*. This follows directly from the observation that a number that falls 2 standard deviations or more away from the mean in its batch is a very unusual number in terms of its batch. Specifically, only about 5% of the numbers in a normally distributed batch fall this far from the mean.

Every error range (or confidence interval) expressed in terms of standard errors corresponds to a specific confidence level. (The terms *confidence interval* and *confidence level* are too close for comfort, considering that they refer to two rather different concepts. Thus the term *error range* is used here in preference to *confidence interval*.) An error range of ±3 standard errors, as illustrated in Fig. 9.8, corresponds to approximately 99.8% confidence. Reading across the row in Table 9.1 that corresponds to 120 $d.f.$, as we did before, and looking for 3 brings us to the next-to-last column, where 3.160 is relatively close to 3. This column is headed 99.8% confidence. Thus when we concluded, on the basis of Fig. 9.8 that it is very likely that our sample comes from a population with a mean of 3.35 cm ± 0.15 cm, that "very likely" actually meant a probability of around 99.8%. There is only about a 0.2% probability that the population our sample came from has a mean less than 3.20 cm or greater than 3.50 cm. Once again, since a normal shape is symmetrical, that means about a 0.1% probability that the population our sample came from has a mean less than 3.20 cm and about a 0.1% probability that it has a mean greater than 3.50 cm.

Finding the confidence level associated with a 1 standard error range is a little more difficult with Table 9.1. Reading across the row for 120 *d.f.*, and looking for 1, we see values that skip from 0.677 to 1.289. The confidence level corresponding to a 1 standard error range thus falls between these two columns. The columns are headed 50% confidence and 80% confidence. For a large sample such as this, the confidence level corresponding to a 1 standard error range actually is about 66%.

ERROR RANGES FOR SPECIFIC CONFIDENCE LEVELS

In some circumstances when we express inferences about population means as error ranges, we simply use 1 standard error as the error range. This has become the normal practice with radiocarbon dates, to choose an example with which archaeologists are comfortable – even those most uneasy about statistics. The error ranges given with radiocarbon dates are understood by convention to be 1 standard error (and we are accustomed to calling them "error ranges" rather than the more statistically traditional "confidence intervals"). These ranges are arrived at by application of precisely the principles we have just discussed to the sample of emitted subatomic particles counted in the laboratory. We can thus apply exactly the same kinds of statements we have just been making about radiocarbon dates. We are moderately confident that the date of death of the carbon atom population from which came the sample that decayed while in the laboratory counter falls within the 1 standard error range specified. More accurately, the probability that the real date of the carbon falls within that range is 66%. This still leaves quite a substantial risk that the real date falls outside that range. If we double the usual range (to arrive at 2 standard errors), we are stating the date less precisely (with an error range twice as large), but we can be 95% confident that the real date falls within that larger range. These ranges, of course, as we have all been warned since we first read introductory textbooks, refer only to the risk of error resulting from the process of measuring the quantity of carbon 14, and are in addition to whatever loss of confidence results from risks like mistaken context, contamination, and the like.

It is worth noting that the standard practice widely accepted by archaeologists in radiocarbon dating is an example of precisely the argument made in Chapter 7 and earlier in this chapter for using samples not strictly randomly selected as a basis for inferences about populations we are interested in. Radiocarbon date error ranges are based on a random sample of carbon 14 atoms (those that decay while the specimen is in the counter). But this sample is drawn from a population of carbon 14 atoms rolled up in an aluminum foil packet in the field through no rigidly random sampling procedure. And the inference made about that population of atoms on the basis of a random sample from it is readily extended to characterize something much broader than that aluminum foil packet. If we are responsible about it, we always bear in mind the risks of mistaken context, contamination, and so on that might invalidate the extension of that inference to the phenomenon we are really interested in, but we do not let the mere existence of such risks paralyze our use of a very powerful dating technique. We can follow just such procedures with many other kinds of samples as well – recognizing the possibility that they may be biased samples from the populations we are really interested in, but that it is worth going ahead and studying them anyway because the possibility of such bias may never be absolutely eliminated.

The 1 standard error range, in any event, has considerable precedent behind it in archaeology because it is the standard for radiocarbon dating. Sometimes an error range of 2 standard errors is used when an author is willing to speak less precisely in exchange for higher levels of confidence. Providing error ranges in this way has one principal disadvantage. The corresponding confidence levels are not entirely self-evident. We found earlier that, in the case of our example sample of 100 projectile points, a 1 standard error range corresponds to about 66% confidence. In the same case a 2 standard error range provides 95% confidence, and a 3 standard error range provides about 99.8% confidence.

These confidence levels can be used as rules of thumb, but they do not hold true if the sample under consideration is small. Suppose our sample had consisted of only six projectile points. We would have needed to use the row in Table 9.1 for 5 $d.f.(n-1)$. In this row, we find a t value of approximately 2 in the column corresponding to 90% confidence rather than the 95% confidence we found before.

To provide error ranges at a fixed level of confidence irrespective of sample size it is necessary to use the t table to determine exactly how many standard errors are required for the desired confidence level. In the case of the sample of 100 projectile points with a mean length of 3.35 cm and a standard deviation of 0.50 cm, we might want to express our estimate of the mean projectile point length in the population with an error range at the 90% confidence level. To do this we find the standard error (as before):

$$SE = \frac{\sigma}{\sqrt{n}} = \frac{0.50\,\text{cm}}{\sqrt{100}} = \frac{0.50\,\text{cm}}{10} = 0.05\,\text{cm}$$

Then we use the t table (Table 9.1) to determine how many standard errors correspond to 90% confidence for a sample of 100. For $n = 100$, $d.f. = 99$, so we use the row for 120 $d.f.$. The value in the column for 90% confidence is 1.658, which means that for a sample of this size an error range of 1.658 standard errors corresponds to a 90% confidence level. We thus multiply the standard error (0.05 cm) by 1.658 to arrive at an error range of ± 0.08 cm. We then say that we are 90% confident that our sample came from a population with a mean of 3.35 cm ± 0.08 cm. If our sample had consisted of 12 projectile points instead of 100, we would have had to use the row in the table for 11 $d.f.$, and we would have needed to use an error range of 1.796 standard errors instead of 1.658. Calibrating error ranges to a specific confidence level in this manner eliminates any possible confusion arising from differing sample sizes, and is generally to be recommended.

Be Careful How You Say It

When you estimate the mean of a population on the basis of a sample and provide an error range for the estimate, it is essential to specify the confidence level as well. Virtually the only exception to this rule is for radiocarbon dates where the convention of providing error ranges of ± 1 standard error is firmly established. The conclusion reached in the example discussed at length in the text might, for example, be stated, "We estimate, on the basis of our sample, that the projectile points used by the inhabitants of our region during the one prehistoric period when the region was occupied had a mean length of 3.35 cm $\pm$ .08 cm (at the 90% confidence level)." Alternatively, we might say, "Our sample indicates 90% confidence that the mean length of projectile points in our region was 3.35 cm $\pm$ .08 cm." It is not incorrect to say, "Our sample indicates 90% confidence that the mean length of projectile points in our region was between 3.27 cm and 3.43 cm." It is probably better, however, to express the error range as a $\pm$ figure associated with the mean. Stating only the maximum and minimum values of the range encourages some people to think that all values within that range are equally likely estimates, and that values outside the range are not possible. We know, however, that the mean itself is the single most likely estimate, and that there is some possibility that the "correct" population value actually lies outside whatever error range is expressed.

FINITE POPULATIONS

The example that we have used throughout this chapter involves a sample selected from a large and vaguely defined population – an infinite population in statistical terms. If the population is small and the sample is a substantial fraction of it, we can take mathematical advantage of an observation that makes intuitive good sense as well. It seems intuitively obvious that, if our sample of 100 projectile points comes from a total population of 120 projectile points, then there is less uncertainty in our estimate of the mean length in the population than if the sample of 100 comes from an effectively infinite population of projectile points. In this case at least, what seems true by common sense can be shown to be true mathematically as well. Whenever the population is finite we can include the *finite population corrector* in the equation for the standard error, thus:

$$SE = \frac{\sigma}{\sqrt{n}}\sqrt{1 - \frac{n}{N}}$$

where σ = the standard deviation in the population (represented by the standard deviation in the sample as before), n = the number of elements in the sample, and N = the number of elements in the population.

This will be recognized as the same equation used before for the standard error with the addition of the term $\left(\sqrt{1 - n/N}\right)$. If the sample is a very large portion of the population, the finite population corrector makes the standard error smaller (hence the error range narrower and precision greater). For example, if we select a sample of 100 from a population of 120, $n = 100$, $N = 120$, and $\sqrt{1 - n/N} = \sqrt{1 - (100/120} = 0.408$. Whatever the standard error would otherwise have been in such an instance, the addition of the finite population corrector makes it only .408 as large (multiplies it by 0.408). On the other hand, if the sample of 100 is selected from a population of 10,000, $n = 100$, $N = 10,000$, and $\sqrt{1 - n/N} = \sqrt{1 - (100/10,000} = 0.99$. Multiplying whatever the standard error would otherwise have been by 0.99 clearly has very little effect on it.

The question arises, then, of when to apply the finite population corrector and when not to. It can always be applied when the number of elements in the population is known. If the population is very large compared to the size of the sample, it will not have much impact on the standard error. If you always use the finite population corrector when N is known, however, it will do its work whenever the sample is a large enough part of the population for it to make a difference. You cannot, of course, apply the finite population corrector when you do not know how many elements are in the population (that is, when the population is, for statistical purposes, infinite).

A COMPLETE EXAMPLE

The discussion of confidence levels and error ranges up to this point has made the whole process seem much more involved and complicated than it really is. This is a consequence of picking the process apart piece by piece to understand why it works the way it does. It is now time to work through an example without all the explanation to show that the procedure of estimating the mean of a population from a sample is really pretty straightforward.

Imagine that we have selected a random sample of 25 bowl rim sherds from the total of 53 bowl rim sherds recovered from a particular house in an excavated village site. We wish to estimate the mean bowl rim diameter in the population of 53 rim sherds on the basis of measurements made on the 25 rim sherds in the sample, and we wish to state our estimate at the 95% confidence level. The measurements are provided in Table 9.2. The stem-and-leaf plot in Table 9.2 confirms that the shape of this batch is roughly single peaked and symmetrical (as least as much as can be expected in a sample this small), so it seems reasonable to use the mean as an index of the center.

The mean of the 25 measurements is 14.79 cm, so the most likely single value for the mean rim diameter in the population of 53 rim sherds is 14.79 cm. The standard deviation in the sample is 3.21 cm so the standard error is

$$SE = \frac{\sigma}{\sqrt{n}} \sqrt{1 - \frac{n}{N}}$$

$$= \frac{3.21 \, \text{cm}}{\sqrt{25}} \sqrt{1 - \frac{25}{53}}$$

Table 9.2. Rim Diameter Measurements for a Sample of 25 Rim Sherds

Diameter (cm)	Stem-and-leaf plot	
7.3		
9.3		
11.6		
11.8	21	0
12.2	20	
12.5	19	45
12.9	18	8
13.3	17	37
13.4	16	25
13.8	15	678
14.0	14	0489
14.4	13	348
14.8	12	259
14.9	11	68
15.6	10	
15.7	9	3
15.8	8	
16.2	7	3
16.5		
17.3		
17.7		
18.8		
19.4		
19.5		
21.0		

$\overline{X} = 14.79\,\text{cm}$

$\sigma = 3.21\,\text{cm}$

$$= \frac{3.21\,\text{cm}}{5}\sqrt{\frac{28}{53}}$$

$$= 0.64\,\text{cm}\sqrt{0.53}$$

$$= 0.47\,\text{cm}$$

Since we need to state our estimate at the 95% confidence level, we must find the value of t corresponding to the 95% confidence level and $n-1$ degrees of freedom. In the row of Table 9.1 for 24 $d.f.$ and the column for 95% confidence, we find the t value 2.064. The error range we state, then, must be 2.064 standard errors. Since the standard error is 0.47 cm, the error range becomes $2.064\,(0.47\,\text{cm}) = 0.97\,\text{cm}$. We can thus state that we are 95% confident that the mean rim diameter for the 53 sherds recovered from this house is $14.79\,\text{cm} \pm 0.97\,\text{cm}$.

HOW LARGE A SAMPLE DO WE NEED?

If we know just what we need to find out before we select a sample, we are in position to determine how large a sample we need in order to achieve our objective. We accomplish this by applying the same reasoning used throughout this chapter, but doing it backward. That is, we decide in advance what confidence level we wish to speak at and how large an error range is acceptable. Then we determine how large a sample will be needed to meet those goals. The one quantity we must guess at is the likely magnitude of the standard deviation in the sample. Such a guess can be difficult to make in practice although it might be based on study of similar known samples.

For example, suppose we wish to estimate the mean thickness of sherds at a site with an error range no more than ± 0.5 mm at a confidence level of 95%. We have measured sherd thicknesses before for collections from a number of sites in the region, and we find that the standard deviation in a sample of sherds is usually somewhere around 0.9 mm. We are willing to take the sherds visible on the surface to represent the sherds present in the site, and we want to send our field assistant to collect a sample of sherds randomly from the surface of the site. So as not to waste time, we would like to say in advance just how large a sample will be necessary. The error range (ER), of course, is t times the standard error, or

$$ER = t \left(\frac{\sigma}{\sqrt{n}} \right)$$

If we solve this formula for n, we get

$$n = \left(\frac{\sigma t}{ER} \right)^2$$

We have previously found the standard deviation in such samples to be about 0.9 mm, so we can use this value for σ. Since we do not yet know the sample size, we will use the row of Table 9.1 for ∞ $d.f.$ to obtain a t value of 1.960 for a 95% confidence level. We want ER to be 0.5 mm. Thus

$$n = \left(\frac{(0.9 \, \text{mm})(1.960)}{0.5 \, \text{mm}} \right)^2$$

$$= \left(\frac{(1.764 \, \text{mm})}{0.5 \, \text{mm}} \right)^2$$

$$= 3.528^2$$

$$12.447$$

We would tell our field assistant to select a sample of 12 or 13 sherds.

To show that this approach works, assume our field assistant returned with a sample of 13 sherds with a mean thickness of 7.3 mm and a standard deviation of

---

The Sample Size, the Sampling Fraction, and Rules of Thumb

The equations we have used in this chapter make clear that sample size is a very important issue. By sample size, statisticians ordinarily mean n, the number of elements in the sample. They do not nearly so often find it useful to think of sample size in terms of the sampling fraction (n/N, the fraction of the population included in the sample). They do not find it useful in the first place because so often samples are drawn from infinite populations (at least ones that are large and not enumerated). If we do not know how many elements are in the population we are sampling from, we clearly cannot begin to say what the sampling fraction is. In the second place, the number of elements in the sample has much greater impact on the results of our calculations than the sampling fraction has. (If you do not believe this, try some experiments with the equations in this chapter, and you will see that it is true.)

This means that when we begin to think about whether a sample is adequate for achieving our aims we must think less in terms of sampling fraction and more in terms of the number of elements in the sample. This shows one of the widespread pieces of conventional wisdom about sampling in archaeology to be a serious misconception. It has often been suggested that a good rule of thumb in sampling is to select a 5% sample. The principles discussed in this chapter make it quite clear that this is *not* a good rule of thumb. Sometimes a 5% sample will be insufficient; other times it will be far more than necessary; if the population is of undetermined size it will be inconceivable.

0.9 mm (as expected). The error range for a 95% confidence level would be

$$ER = t \left(\frac{\sigma}{\sqrt{n}} \right)$$

With a sample size of 13, we find that t for 12 $d.f.$ and 95% confidence is 2.179, so

$$ER = 2.179 \left(\frac{0.9\,\text{mm}}{\sqrt{13}} \right)$$
$$= 2.179 \left(\frac{0.9\,\text{mm}}{3.606} \right)$$
$$= 2.179 \,(.250\,\text{mm})$$
$$= 0.54\,\text{mm}$$

Thus we would conclude that the mean thickness of sherds at the site in question is 7.3 mm ± 0.5 mm at the 95% confidence level. We have achieved our goal of estimating the thickness with an error range of no more than about 0.5 mm at the 95% confidence level.

Thinking about the confidence and precision we need in making specific estimates is one sound way to approach the always vexing question of how large a sample is needed. Following this approach, of course, requires deciding specifically what we want to find out, how precise our results need to be, and how confident we want to be of our conclusions. These parameters are not absolutes. They vary from one situation to the next. What is sufficient precision in one context may be hopelessly imprecise in another. And what is sufficient confidence for some purposes may be altogether inadequate for others. If we cannot state our aims clearly enough to at least approximate how large a sample may be needed to achieve them, however, it is probably premature to be selecting a sample. We should go back and think harder about exactly what we are trying to find out.

ASSUMPTIONS AND ROBUST METHODS

The use of most of the tools discussed in this and subsequent chapters requires making some assumptions. These will be discussed at the close of each chapter. Most of the techniques are already fairly *robust*. That is, they can be applied to samples that only approximately meet the assumptions. And there are things we can do even with samples that violate the assumptions drastically.

Once we have decided that we are willing to treat a batch of numbers as a random sample from a larger population we wish to know about, the only assumption we must make in order to estimate the population mean and attach error ranges to it in the manner described here is that the special batch must have an approximately normal distribution. The central limit theorem tells us that this will always be the case for large samples (that is, larger than 30 or 40 elements). When working with a smaller sample, it is wise to look at the stem-and-leaf plot to check for a roughly symmetrical and single-peaked shape. If a small sample has a single-peaked and roughly symmetrical shape, then we can count on its special batch to have a normal shape. If a small sample has a badly skewed shape we might try to correct this with transformations, but this is not very useful for estimating means because we would wind up estimating something like the mean of the logarithm of the measurement in the population, and such a quantity is not very easy to relate to what we want to know.

Looking at a stem-and-leaf plot should always be the initial step anyway, even with a large sample. This is because the sample might have outliers or a badly skewed shape that would make the mean and standard deviation meaningless as numerical indexes of level and spread, as discussed in Chapters 2 and 3. If a sample has outliers or a badly skewed shape, then the population the sample was selected from probably does too. In such a case, the mean will likely not be a good index of center for the population, and is thus not what we want to estimate. If the problem is outliers, the trimmed mean is a better index of the center. If the problem is skewness, then the median is a better index of the center. In such cases, it makes sense to estimate, not the regular mean of the population, but the trimmed mean or the median of the population instead. Estimating the trimmed mean is dealt with below because it is a natural extension of estimating the mean, which we have just discussed. Estimating the median requires such a different approach that it is left for a separate chapter.

The best estimate of the trimmed mean of the population is simply the trimmed mean of the sample. Error ranges for different confidence levels can be provided for this estimate of the trimmed mean of the population following exactly the same procedures used to provide error ranges for estimates of the regular mean. The only difference is that, instead of using the sample size, the mean, and the standard deviation, their values are replaced in all equations with the trimmed sample size, the trimmed mean, and the trimmed standard deviation. Otherwise, everything about the calculations remains the same.

Table 9.3 lists a small sample of projectile point weights. The stem-and-leaf plot shows upward skewness and/or high outliers. The mean of this sample is 47.45 g, which falls too far above the center of the principal bunch of values to be a very useful index. If the sample is like this, the population probably is too. The trimmed mean would be a much more meaningful index of the center of such a shape. A 15% trimming fraction would eliminate the three high outliers which are causing most of the difficulty. The 15% trimmed mean, then, is 37.9 g, which falls where an index of the center of this batch should fall. The variance of the Winsorized batch

Table 9.3. Weights of a Small Sample of Projectile Points

Weight (g)	Stem-and-leaf plot	
96		
37	15	6
28	14	
34	13	
52	12	
18	11	
21	10	8
39	9	6
156	8	
43	7	
44	6	
19	5	25
30	4	347
108	3	014799
55	2	1488
24	1	89
28		
47		
39		
31		

Be Careful How You Say It

If you estimate the trimmed mean for a population rather than the regular mean, you must make it very clear what you've done. Be sure to refer to what you've estimated as the "trimmed mean," never just the "mean," and specify the trimming fraction as well. Just as with estimates of the regular mean, the confidence level for which the error range was calculated must be given too. For the example in the text, we might say, "On the basis of our sample, we estimate that the 15% trimmed mean weight of projectile points is $37.9\,g \pm 8.2\,g$ at the 95% confidence level."

---

is 137.19, so the trimmed standard deviation is 14.16 (see Chapter 3). The standard error, then, is

$$SE = \frac{\sigma}{\sqrt{n}} = \frac{14.16}{\sqrt{14}} = 3.8\,g$$

For an error range at the 95% confidence level, we would multiply the standard error by the value of t for 13 $d.f.$ $(n_T - 1)$. The 95% confidence error range, then, is $\pm 8.2\,g$ (that is, 3.8×2.160). Estimating the trimmed mean instead of the regular mean for this population is not only more meaningful (it avoids the effects of the high outliers) but also more precise. The error range for 95% confidence that we would have to provide for an estimate of the regular mean would be $\pm 16.1\,g$. This is because the outliers that are eliminated by trimming would cause the regular standard deviation of the sample to be quite large. Consequently the standard error and the 95% confidence error range would be quite large as well. Estimating the trimmed mean, then, pays off double in this instance – it is a more sensible index of the center for these numbers, and its estimate comes with a much smaller error range.

PRACTICE

1. You have tested a newly reported neolithic site at Châteauneuf-sur-Loire. You decide that you are justified in working with the artifacts from your test pits as if they were a random sample of the utilized flakes in the site. The lengths of the flakes are given in Table 9.4. Estimate an appropriate numerical index of center for length of utilized flakes in the site on the basis of this sample. Provide an error range for this estimate at the 95% confidence level. State in one clear sentence what this estimate and its error range mean.

2. You decide that the estimate that you have made for utilized flake length at Châteauneuf-sur-Loire is not precise enough. You would like an estimate with

**Table 9.4. Lengths (in cm) of 40 Utilized Flakes
from Châteauneuf-sur-Loire**

4.7	6.8	3.5	5.9	6.5
4.1	6.2	6.0	7.8	8.8
8.0	9.3	8.3	8.1	7.4
3.2	6.9	5.5	4.3	8.5
9.7	7.3	4.3	4.7	6.3
7.5	4.5	4.8	3.0	7.0
5.7	3.9	5.6	6.1	5.3
5.0	5.4	6.1	5.1	2.6

**Table 9.5. Diameters (in m) of 44 Mesolithic
Hearths at Berwick-upon-Tweed**

0.91	0.75	1.03	0.82	2.13
0.51	0.80	0.66	0.93	0.66
0.76	0.90	0.76	0.95	0.62
1.64	0.58	0.96	0.56	1.93
0.85	0.60	0.74	0.78	0.68
0.88	0.70	0.64	0.89	0.80
0.72	2.47	0.62	0.98	0.74
0.77	0.84	0.86	1.08	0.93
0.69	1.00	0.84	0.83	

**Table 9.6. Zinc (in Parts Per Million) for 14 Obsidian
Blades from a Prehistoric House at Huancabamba**

53	49	41	59	74
37	66	33	48	57
60	55	82	22	

an error range for 95% confidence that is only half as large as the one you just calculated, so you return to the site for more fieldwork in order to obtain a larger sample. How large a sample of utilized flakes will you need to achieve your aim?
3. You have excavated a mesolithic site at Berwick-upon-Tweed and found a remarkable number of well-formed hearths. Their diameters are given in Table 9.5. Using this set of hearths as a random sample of hearths at the site, estimate an appropriate numerical index of center for hearth diameters at the site as a whole. Provide an error range for this estimate at the 99% confidence level.
4. You have excavated the complete and well-preserved remains of a single prehistoric household at Huancabamba, and the artifacts recovered include 37 obsidian blades. In order to compare this assemblage with others and with different obsidian raw material sources, you wish to know the mean zinc content in the chemical composition of these 37 blades. Since zinc occurs in very small amounts, it is quite expensive to measure, so even though the entire assemblage is small, you

treat it as a population from which you select a random sample of 14 blades to analyze. The quantity of zinc found in each blade (in parts per million) is given in Table 9.6. Estimate the mean number of parts per million of zinc in the population of 37 blades. Provide an error range for your estimate at the 90% confidence level. State the meaning of this estimate and its error range in a single clearly constructed sentence.

Chapter 10
Medians and Resampling

Classical statistical theory provides powerful tools for estimating population means from samples and for establishing error ranges for desired confidence levels, and these were presented in Chapters 8 and 9. If outliers in a sample interfere with using the mean, the same tools can be applied to estimate the trimmed mean. If the asymmetrical shape of a sample (skewness) interferes with using the mean, trans-formations can be applied as a correction, and this makes it possible to proceed with significance testing, as we will see in Chapters 11–15. While this works fine for sig-nificance testing, it puts the measurements on a scale that is not intuitive and makes it difficult to talk about them straightforwardly. It would just not be at all easy to talk meaningfully about estimates of, say, the mean logarithm of site area in two periods. The median may be a more useful index of center in such a case, and the best estimate of the median in a population is the median in the sample. There is, however, no abstract theoretical basis for establishing error ranges for this estimated median at any particular confidence level because there is no theoretical way to determine the center, spread, or shape of the special batch (or sampling distribution) of the median as there is for the mean. The contribution of exploratory data analy-sis to this difficulty was to recognize that the special batch can be approximated by *resampling*, or repeatedly selecting samples from the sample itself.

The back-to-back stem-and-leaf plot of Early and Late Classic period site areas in Table 10.1 provides a prime example. The 113 Early Classic site areas range from less than 1 ha to 211 ha, and the 95 Late Classic ones from less than 1 ha to 101 ha. As is often the case with site areas, these batches straggle upward, and some of the higher values in at least the Early Classic batch might well be identified as outliers. Although the very largest sites occurred in the Early Classic, if we focus on the main bunch of numbers we see that in general it is Late Classic sites that are somewhat larger. The change is not dramatic, but it shows in the box-and-dot plot in Fig. 10.1. The median, the lower and upper quartiles, and the extreme adjacent values are all higher for the Late Classic. High outliers, however, are less numerous

R.D. Drennan, *Statistics for Archaeologists*, Interdisciplinary Contributions
to Archaeology, DOI 10.1007/978-1-4419-0413-3_10,
© Springer Science+Business Media, LLC 2004, 2009

**Table 10.1. Back-to-Back
Stem-and-Leaf Plot of Early and Late
Classic Period Site Areas**

Early		Late
1	21	
	20	
	20	
	19	
	19	
	18	
	18	
	17	
	17	
	16	
	16	
	15	
4	15	
	14	
	14	
	13	
0	13	
	12	
	12	
	11	
4	11	
9	10	
	10	1
	9	59
03	9	3
	8	
33	8	
9	7	556
	7	34
5557	6	588
2234	6	000
7889	5	5667
00244	5	00123
567899	4	566889
01112233	4	0012233
566789	3	556679
00112344	3	001344
555567889	2	55667889
02222344	2	00112234
55567788	1	55677788999
001111112234	1	0122344
566666677888999	0	67888899
001134	0	0344

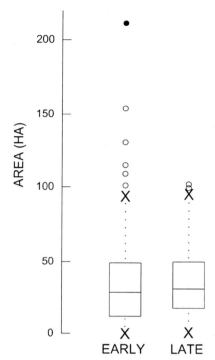

Figure 10.1. Box-and-dot plots comparing Early and Late Classic period site areas.

and less extreme in the Late Classic. These are exactly the circumstances in which the mean is likely to be misleading, and this is indeed the case. The mean site area for Early Classic is 36.3 ha; for Late Classic, it is 35.7 ha, suggesting that the center of the batch has shifted down, not up. For both batches, the mean is higher on the scale than the visible center of the batch in the stem-and-leaf plot. Just as we might expect, the median provides a better index of center for both batches. The median site area for Early Classic is 28.2 ha; for Late Classic, it is 30.6 ha, showing the upward shift that we see in the stem-and-leaf and box-and-dot plots. For summing up the comparison, then, it is satisfying to say that the median site area has increased from 28.2 ha to 30.6 ha.

If these two batches of measurements are to be taken as samples from the populations we really need to talk about, however, we may want to make estimates for the population with error ranges at a particular confidence level. As we have seen, using the mean is unsatisfactory. The trimmed mean would be an improvement, especially in removing the numerous outliers from the Early Classic batch, but the real problem would still remain, since it concerns the fundamental shape of both batches, which is skewed upward. Transformations would make significance testing possible, but it would lead to an unwieldy comparison of, say, the negative reciprocals of site areas between the two periods. Resampling makes it possible to put error ranges with the medians in a case like this.

THE BOOTSTRAP

The most commonly used resampling technique is the *bootstrap*. It consists of taking the sample batch as if it were a population and repeatedly selecting new samples from the sample. The new samples are selected randomly and typically are samples of the same size as the original sample batch. They differ from the original sample batch only in that sampling *with replacement* will produce new samples that vary by randomly omitting different cases and including other cases multiple times. In performing the bootstrap, at least 1,000 resamples are usually selected. The median of each resample is found, and a batch accumulates that consists of the medians of all the resamples. This batch can be used to accomplish the same thing that the special batch makes it possible to do for means. That is, it can be treated as the sampling distribution of the median.

When estimating the mean, we know that the special batch has a normal shape (as long as the sample is more than 30 or 40), and we can calculate its mean and its standard deviation. Then, with the mean and standard deviation of the special batch, we can figure out how unusual it would be to get a sample like the one we have from a population with a mean rather different from the mean of our sample, as we saw in Chapters 8 and 9. The special batch for the median, however, consisting of the medians of all the resamples cannot be counted upon to be single peaked and symmetrical. In fact, for medians, it is almost always very asymmetrical and often has multiple peaks. The histogram in Fig. 10.2, for example, shows the distinctly two-peaked and asymmetrical shape of the batch consisting of the medians from

Figure 10.2. Histogram of the site area medians for the 10,000 resamples from the Early Classic period sample.

10,000 resamples from the Early Classic site area batch from Table 10.1. The mean and standard deviation would be poor indexes of center and spread for this batch, so they would not provide us with a useful approach to unusualness within the batch. The median of this batch of 10,000 resample medians, though, is 28.2 ha, the same as the median of the original sample and a good index of the center of the special batch as well.

We saw in Chapter 4 that percentiles, familiar to students from the reports of standardized tests, are a way of characterizing unusualness, and it is percentiles that provide the most useful way to approach unusualness in a very non-normal batch like the one in Fig 10.2. This special batch can be taken to represent the set of medians of populations our sample might have come from. In order to find an error range for, say, a 90% confidence level to attach to the median of 28.2 ha, we would look in this batch of 10,000 resample medians for the 5th and 95th percentiles. That is, the middle 90% of resample medians would represent the range within which we would be 90% confident that the median of the population lies. We would, then, want to find the number below which 5% of the medians fall, and the number above which 5% of the medians fall, leaving 90% of the resample medians between these two numbers. Since 5% of the 10,000 medians would be 500, we would want the 500th and 9,500th numbers in the batch (either counting up from the lowest or down from the highest). For this special batch these two numbers are 24.6 and 35.0 ha.

Finally, then, we would estimate the median site area for the innumerably large population of all Early Classic sites in our region as 28.2 ha. And we would be 90% confident that the median in this population lies between 24.6 ha and 35.0 ha. As is usual with bootstrapped error ranges for the median, the error range is not symmetrical. It runs from 3.6 ha below the median of 28.2 ha to 6.8 ha above it and thus cannot be expressed as a ± figure. An error range for any particular confidence level can be determined by selecting appropriate percentiles. An error range for the 95% confidence level lies between the 2.5th and 97.5th percentiles; for the 98% confidence level, between the 1st and 99th percentile; and for the 99% confidence level, between the 0.5th and 99.5th percentile.

Statpacks

Resampling approaches like the bootstrap have been somewhat slow to appear in statpacks, but their presence is getting more common. Finding an error range for the median with the bootstrap is still likely, however, to involve more than simply selecting that option from a single menu. It may involve choosing an option to perform resampling, selecting the bootstrap as the resampling technique to be used, setting how many resamples are to be chosen (usually at least 1,000), and specifying that the median is the desired statistic. The statpack is then likely to save the medians from all the resamples in a new data file, within which you will need to find the appropriate percentiles to establish the size of the error range for the desired confidence level.

The bootstrap may seem as magical a notion as pulling yourself up by your bootstraps, and that is exactly how it got its name. It does not derive from abstract mathematical logic. Repeated experimentation, however, has shown that the bootstrap provides a very good assessment of error ranges for different confidence levels. It can be used to find error ranges for means, too, even though the classic theoretically derived approach makes this unnecessary. For example, the mean of the batch of Early Classic site areas is not a very good index of center, as noted earlier, but it can be estimated for the population this sample comes. If we do this by the standard approach discussed in Chapter 9, we estimate that the mean Early Classic site area in the population is 36.3 ha ± 60 ha (at the 95% confidence level).

If we make this estimate by bootstrapping, we produce a batch consisting of the means of 10,000 resamples. This special batch is single peaked and symmetrical, just as the central limit theorem tells us the special batch of the mean should be for a sample this large. The mean of the 10,000 resample means is 36.3 ha, providing us with exactly the same estimated mean for the population as classical theory did. Since the batch is single peaked and symmetrical, we can use the mean and standard deviation to deal with unusualness, and again we arrive at an error range of ±6.0 ha for the 95% confidence level. The two results will not always agree this perfectly. Rounding error and other factors will produce slight variations if the calculations are carried out to enough decimal digits of precision, just as will happen if the exact same calculation is done on calculators operating at different levels of precision.

In addition to the bootstrap, there is a second resampling approach, called the *jackknife*. The jackknife is just like the bootstrap except that the resamples are selected slightly differently. Instead of selecting, with replacement, a large number of resamples the same size as the original sample, jackknife resamples are produced by omitting each case from the original sample in turn, one by one. Thus the resamples are smaller by one case than the original sample, and there are only as many of them as there are cases in the original sample. The jackknife is somewhat less robust than the bootstrap and less often used.

PRACTICE

1. Look back at the data on Late Bronze Age sites from Nanxiong in Table 3.5. In the practice questions there, you will have recognized that this batch is asymmetrical and not suitably characterized by the mean. The median, however, provides a good index of center for it. Treat this batch of site areas as a sample from a larger population of Late Bronze Age sites, and estimate the median site area in that population. Use the bootstrap to provide an error range for your estimate at the 90% confidence level. State in one clear sentence what this estimate and its error range mean.

Chapter 11
Categories and Population Proportions

Chapters 7–10 dealt with making estimates about a population on the basis of a sample when the observation of interest was a measurement whose mean or median in the population we wished to estimate. In Chapter 6 we discussed a different kind of observation, one based on categories rather than measurements. If the observation of interest involves a set of categories rather than a measurement, it of course makes no sense to think in terms of the center of a batch or its spread. Rather, we approach the batch in terms of proportions. When we observe categories in a sample, then, our basic thought about the population from which the sample was selected concerns the proportions of the different categories in the population, not the mean or median of anything.

The estimation of a population proportion on the basis of a sample is quite similar to the estimation of a population mean on the basis of a sample, so in this chapter we will treat proportions as an extension of the principles applied to means in the previous three chapters. Suppose that we examine the raw materials used to manufacture the projectile points in the sample of 100 projectile points discussed in Chapter 9. We may find that, of the 100 points, 13 are made of obsidian. Since the number in the sample is 100, the proportion of points made of obsidian in the sample is 13/100 or 13.0%. What does this tell us about the large and vaguely defined population that the sample of 100 points came from? Just as with means, the sample proportion is the likeliest single value for the proportion in the population from which the sample was selected. Thus, the best estimate of the population proportion, based on this sample, is 13.0%.

Just as it is possible that a sample may have a different mean than the population it came from, it is possible to select a sample with a proportion of 13.0% obsidian projectile points from a population with a proportion of obsidian projectile points different from 13.0%. Thus we would like to attach an error range for a given

R.D. Drennan, *Statistics for Archaeologists*, Interdisciplinary Contributions
to Archaeology, DOI 10.1007/978-1-4419-0413-3_11,
© Springer Science+Business Media, LLC 2004, 2009

confidence level to this estimate just as we did to estimates of population means. We can use the standard error for this purpose in the case of proportions as well. The only difficulty is that calculation of the standard error of the mean was based on the standard deviation in the sample, and there is no obvious intuitive meaning to the concept of the standard deviation of a proportion. It can be shown mathematically, however, that there is a very simple equivalent of the standard deviation for proportions:

$$s = \sqrt{pq}$$

where s = standard deviation of the proportion, p = the proportion expressed as a decimal fraction, and $q = 1 - p$.

In our example, the proportion of obsidian projectile points in the sample, expressed as a decimal fraction, is 0.130 and $q = 1 - p = 1 - 0.130 = 0.870$. Thus

$$s = \sqrt{pq} = \sqrt{(0.130)(0.870)} = \sqrt{0.1131} = 0.3363$$

This standard deviation of a proportion does connect in a commonsense way with the standard deviation of the mean. We know that a small standard deviation indicates a batch with a small spread, and that such a batch can also be called highly homogeneous. A homogeneous batch with regard to proportions would be a batch with a very large (or very small) proportion of the category of interest. If 99% of the projectile points were made of obsidian, this very homogeneous batch would have a low standard deviation: $\sqrt{pq} = \sqrt{(0.99)(0.01)} = \sqrt{0.0099} = 0.0994$. A batch with 1% obsidian and 99% nonobsidian projectile points would, of course, yield the same result. The most heterogeneous possible batch in this regard would have 50% obsidian projectile points, and its standard error would be $\sqrt{pq} = \sqrt{(0.50)(0.50)} = \sqrt{0.2500} = 0.50$. The more heterogeneous batch thus has the larger standard deviation, just as it should.

This standard deviation is used in calculating the standard error by exactly the same procedure used for means. Since σ, the population standard deviation, is unknown, we use the sample standard deviation, s, in the equation

$$SE = \frac{\sigma}{\sqrt{n}} = \frac{0.3363}{\sqrt{100}} = \frac{0.3363}{10} = 0.03363$$

The standard error of the proportion in our example, then, is 0.034 or 3.4%. We can use this as a 1 standard error range attached to the estimated proportion and say that the population proportion is 13.0% ± 3.4%, or between 9.6% and 16.4%. As usual with a 1 standard error range, we would be about 66% confident that the proportion in the population our sample was selected from fell between 9.6% and 16.4%.

To adjust the error range thus obtained to the specifically desired confidence level, we would use Student's t distribution (Table 9.1) to determine t for the given number of degrees of freedom and confidence level and multiply the standard error by that value. To adjust the error range in this example to a 95% confidence level, we would use the row in Table 9.1 for 120 $d.f.$ (the closest available to $n - 1 = 99$ $d.f.$) and find in the 95% confidence column that $t = 1.98$. Multiplying the standard error by 1.98 yields $(0.034)(1.98) = 0.067$. Thus, at a 95% level of confidence, we would estimate that the proportion of obsidian projectile points in the population from which our sample was selected is 13.0% $\pm$ 6.7% (or between 5.3% and 19.7%). This means, of course, that there is only a 5% chance of selecting a sample like ours (that is, a sample of 100 with a proportion of 13.0% obsidian projectile points) from a population with a proportion of obsidian projectile points less than 5.3% or greater than 19.7%.

The finite population corrector can be applied to the calculation of the standard error of a proportion just as with a mean. For example, suppose that in the complete excavation of a village site occupied for a relatively short period of time, we identify the remains of 24 houses. In the cases of 17 of the 24 houses, the remains are well enough preserved to enable us to determine the locations of the entrances. Of these 17 houses, 6 had their entrances facing south. After careful consideration of possible sources of bias, we decide that we will treat the 17 houses as a random sample from the population of 24 houses originally built at the site. We thus estimate that 6/17, or 35.3%, of the houses at the site had their entrances facing south. The standard error of this proportion will be

$$SE = \frac{\sigma}{\sqrt{n}}\sqrt{1 - \frac{n}{N}}$$

where $\sigma = s = \sqrt{pq}$.

Thus,

$$\begin{aligned}
SE &= \frac{\sqrt{pq}}{\sqrt{n}}\sqrt{1 - \frac{n}{N}} = \sqrt{\frac{pq}{n}\left(1 - \frac{n}{N}\right)} \\
&= \sqrt{\left(\frac{(0.353)(0.647)}{17}\right)\left(1 - \frac{17}{24}\right)} \\
&= \sqrt{(0.0134)(1 - 0.7083)} \\
&= 0.0625
\end{aligned}$$

If we wish to speak at a 90% confidence level, then we multiply this standard error by 1.746 (t for 90% confidence and 16 $d.f.$ is 1.746 according to Table 9.1) to get an error range at the 90% confidence level of 0.1091. We can thus conclude that, of the 24 houses at the site, 35.3% $\pm$ 10.9% (or between 24.4% and 46.2%) had their entrances facing south. Since this is a finite population we can also convert this estimated proportion (and its attached error range) into numbers of houses for the entire population. Multiplying the lower extreme of the error range (24.4%) by the number of houses in the population (24) gives us 5.9 houses, and multiplying the upper extreme of the error range (46.2%) by the number of houses in the population (24) gives us 11.1 houses. Thus we can say that we are 90% confident that some 6 to 11 houses at the site had their entrances facing south.

In this example, the sample – and, for that matter, the population from which it was selected – is so small that these statistical results may not seem very helpful.

After all, we already knew there were at least six houses with their entrances facing south; there were 6 known south-facing entrances in the sample. And we knew there could not be more than 13 south-facing entrances. There were only 7 houses whose entrances were undocumented. If they all faced south, together with the 6 in the sample, that would make 13. If we already knew that the number of houses with south-facing entrances had to be between 6 and 13, what have we gained by saying that we have 90% confidence that the number of houses with south-facing entrances at this site is between 6 and 11? More than anything else, we have gained an awareness that our sample is quite small for saying anything very precise about the overall population with much confidence. For at least some purposes, this sample would simply be too small to tell us what we need to know, even though it was a 71% sample. (A sample of 17 houses represents 71% of the population of 24 houses.) A sample of 17 is, in statistical terms, a very small sample, no matter how large a proportion of the population it is. If we are working with a sample this small, there is an uncomfortably large risk that whatever proportions we find in it may be quite different from the proportions in the population from which it was selected. Whatever conclusions we derive from this sample about the population from which it was selected cannot be terribly precise or certain, even though they still do constitute our best guess about the population as a whole. Calculation of an error range for a specified confidence level, in this case, tells us that our best guess really is not very good, and *that* is important to know before we go on to use this observation as evidence for or against someone's theory.

HOW LARGE A SAMPLE DO WE NEED?

We can also put such knowledge to use in considering in advance roughly how large a sample we may need for a particular purpose, just as we can when estimating population means. The equation is the same:

$$n = \left(\frac{\sigma t}{ER} \right)^2$$

and we use $\sqrt{pq}$ for σ just as we did above. For example, suppose we wish to know how large a random sample of sherds we must collect from a site in order to estimate the proportions of the various pottery types in its ceramic assemblage with error ranges no wider than ±5% at the 95% confidence level. We must make some guess at the proportions we may actually need to estimate in order to arrive at σ. If we have no idea, then we can use the most conservative guess of 50% since error ranges turn out to be widest when the proportion is 50%. To the extent that the actual proportions we get differ from 50%, then the error ranges will be even narrower than we require. Using 50%, $\sqrt{pq} = \sqrt{(0.50)(0.50)} = 0.50$, so we use 0.50 for σ. The value of t for 95% confidence and ∞ $d.f.$ (since we do not yet know what n will be) is 1.96. Thus

$$n = \left(\frac{(0.50)(1.96)}{0.05} \right)^2 = 384.16$$

We should, then, collect a random sample of some 384 sherds.

If we do so, and discover that 192 of the sherds are of a particular ceramic type, then that type represents 192/384 or 50.0% of the sample. We estimate that the type comprises 50.0% of the total ceramic assemblage at the site (the population from which the sample was selected). The standard error of this proportion is

$$SE = \frac{\sigma}{\sqrt{n}}$$

and we use $\sqrt{pq}$ for σ. Thus

$$SE = \frac{\sqrt{(0.50)(0.50)}}{\sqrt{384}} \quad \frac{0.50}{19.5959} = 0.0255$$

For an error range at the 95% confidence level, we multiply this standard error by the value of t for 95% confidence and ∞ $d.f.$, since 383 $d.f.$ falls far beyond 120, the last row of Table 9.1 before ∞. The 95% confidence level error range, then, is 1.96 standard errors: 0.050 or 5.0%. Thus we estimate that this ceramic type composes 50.0% ± 5.0% of the sherds at the site, and we have achieved the level of confidence and the precision that we required of our sample.

If another pottery type was represented by only 14 sherds in the sample, then we would estimate that it makes up 3.6% of the sherds at the site. In this case we would achieve greater precision at the same level of confidence because the standard error for this smaller proportion would be smaller:

$$SE = \frac{\sqrt{(0.036)(0.964)}}{\sqrt{384}} = \frac{0.1863}{19.5959} = 0.0095$$

Multiplying this standard error of 0.0095 by t for ∞ $d.f.$ and 95% confidence yields $(0.0095)(1.96) = 0.019$ or 1.9%. We could conclude with 95% confidence that this second pottery type represented 3.6% ± 1.9% of the ceramic assemblage at the site.

The difficulties of outliers and asymmetrical shapes that sometimes pose problems in the analysis of measurements and in the estimation of population means simply do not arise with categories and the estimation of population proportions. Thus it is not necessary to consider robust methods here.

PRACTICE

1. In systematic surface collection at the site of Mugombazi you recovered 342 flaked stone artifacts. After careful consideration of possible sources of sampling bias, you decide you will take these 342 as a random sample of the flaked stone

in the site. Of the 342 flaked stone artifacts in the sample, 55 are identified as gravers. Estimate the proportion of gravers in the flaked stone assemblage at the site. Provide an error range for your estimate at the 99% confidence level. In one clearly constructed sentence, express this estimate, providing all the information your reader would need to know to make full use of it.

2. Not far south of Mugombazi lies another extremely large lithic scatter at Bwana Mkubwa. You intend to make a surface collection in such a manner as to have a random sample of the flaked stone at the site. Your aim is to estimate the proportions of different categories of flaked stone artifacts in the overall flaked stone assemblage, and you want estimates for which the error ranges (at a 90% confidence level) are never more than ±5%. How large a sample of artifacts should you select?

3. You proceed to Bwana Mkubwa and make the surface collection as planned (except, of course, for the incident with the rhinoceros). To keep your mind off the pain, and to kill time while waiting in the emergency room, you have an initial look at the artifacts. It turns out that fully 45% of the flaked stone in the sample consists of debitage. Estimate the proportion of debitage in the flaked stone of the site as a whole. Provide an error range for your estimate at a 90% confidence level. State your results in a single clear sentence. Is the error range at least as small as the ±5% you wanted? If not, go back, figure out what went wrong, and try again. (This time, please watch out for the wildlife.)

Part III
Relationships between Two Variables

Chapter 12
Comparing Two Sample Means

Up to now we have concentrated on single batches of numbers and on using single batches of numbers as samples for purposes of making inferences about the populations from which they were selected. The principles discussed in Chapters 7–11, however, can also be applied to the task of comparing batches, which we began to explore in Chapter 4.

Figure 12.1 compares two batches of numbers. These batches are the areas (in square meters) of house floors for two periods (Formative and Classic). After careful consideration of possible sources of bias we decide to work with these two batches as if they were random samples, taking each as a sample of house floors for its period. The sample for the Formative period consists of 32 house floors, and the sample for the Classic period consists of 52 house floors. We begin to explore the two samples with a back-to-back stem-and-leaf plot at the left in Fig. 12.1. This plot reveals that both samples are single peaked and symmetrical enough that the mean would be a useful index of their levels. Neither is a perfect single-peaked and symmetrical shape, but both are quite as single peaked and symmetrical as one has any right to expect in relatively small batches of numbers like this.

The impression gained from the back-to-back stem-and-leaf plot is confirmed by the box-and-dot plot in the center of Fig. 12.1. The box-and-dot plot in addition provides a clear view of the fact that the center of the sample of house floors from the Classic period is higher than the center of the sample of floors from the Formative period. That is, Classic period houses, with a median area of $26.3\,\mathrm{m}^2$, were in general somewhat larger than Formative period ones, with a median area of $24.3\,\mathrm{m}^2$. (This remains a useful thing to say, even though there is considerable overlap in house size between the two periods and even though the smallest house of all dates to the Classic period – both of which facts are evident in the stem-and-leaf plot and

R.D. Drennan, *Statistics for Archaeologists*, Interdisciplinary Contributions
to Archaeology, DOI 10.1007/978-1-4419-0413-3_12,
© Springer Science+Business Media, LLC 2004, 2009

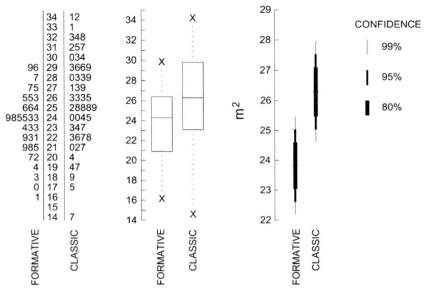

Figure 12.1. Comparison of Formative and Classic period house floor areas (in square meters).

Table 12.1. Comparison of Formative and Classic Period House Floor Samples

	Formative	Classic
$n =$	32 floors	52 floors
$Md =$	$24.3\,m^2$	$26.3\,m^2$
$\overline{X} =$	$23.8\,m^2$	$26.3\,m^2$
Midspread $=$	$4.1\,m^2$	$6.7\,m^2$
$s =$	$3.4\,m^2$	$4.5\,m^2$
$SE =$	$0.60\,m^2$	$0.63\,m^2$

in the box-and-dot plot.) The Classic period sample has a slightly larger spread than the Formative period sample does, although the two samples are not too different in this regard.

Table 12.1 provides the specific figures that compare the two samples in terms of level and spread. Whether we compare medians or means, Classic period house floors seem somewhat larger. And whether we compare midspreads or standard deviations, Classic house floor areas show a slightly larger spread.

Combining these observations (of the sort we made in comparing batches in Chapter 4) with what we know of the behavior of random samples (see Chapters 7–9) might lead us to wonder whether the differences we observe between these two batches are "real" or whether they are just the result of the simple fact that samples do not always very accurately represent the population from which they were selected. We know that if we selected a number of random samples from exactly the same population we would get considerable variation from one to the next. Such random variation between samples is often referred to as the *vagaries of sampling*. In comparing Formative and Classic period house floor areas, might we be seeing nothing more than this kind of random variation between samples? We know, of course, that these two samples did not actually come from the same population – one is a sample from Formative period house floors, and the other is from Classic period house floors. We often say that we imagine, though, that two such samples might have come from the same population, by which we really mean that the two samples might have come from two populations that have the same mean. If our two samples actually came from two populations with identical means, then the mean area of Formative and Classic house floors was the same. Since the means of our two samples are different, we would certainly guess that the means of the populations they were drawn from were different. Nevertheless, based on our previous discussion of the behavior of random samples, we recognize that there is some possibility that both samples could have been drawn from populations with means of, say, $25.0\,m^2$. If this were actually the case, then we would attribute the differences we observe in our samples of Formative and Classic period house floors to the vagaries of sampling. We would not take them to indicate a change in house floor area between the Formative and Classic periods. In Chapter 9 we dealt with this sort of question for one sample at a time by establishing error ranges for various confidence levels, but now we have two samples, which makes the situation more complicated. We can, nevertheless, approach the question in exactly the same way, taking each sample and its parent population in turn.

Table 12.2 provides estimates of Formative and Classic period house floor size for three different confidence levels. These estimates and their attached error ranges were calculated following exactly the procedures presented in Chapter 9. The two samples were treated independently, and error ranges for 80% confidence, 95% confidence, and 99% confidence were calculated separately on the basis of each sample. These error ranges are presented graphically at the right in Fig. 12.1. Such graphs can be referred to as *bullet graphs* (because the representation of error ranges looks ever so slightly like a bullet). A bullet graph makes it easy to compare the two samples not only in terms of their means but also in terms of the confidence implications of their various error ranges. The thickest *error bar* represents the error range for the 80% confidence level. This is the most precise estimate and, correspondingly, the one in which our confidence is lowest. The medium thickness error bar represents the error range for the 95% confidence level. This error range is wider, but our confidence in this less precise estimate is higher. Finally, the thinnest error bar

Table 12.2. Comparison of Formative and Classic Period House Floor Samples

	Mean area	
Confidence level	Formative	Classic
80%	$23.8 \pm 0.8\,m^2$	$26.3 \pm 0.8\,m^2$
95%	$23.8 \pm 1.2\,m^2$	$26.3 \pm 1.3\,m^2$
99%	$23.8 \pm 1.6\,m^2$	$26.3 \pm 1.7\,m^2$

represents the 99% confidence level error range, still less precise and thus worthy of still higher confidence.

Note that this is not simply a different way of drawing a box-and-dot plot, although both bullet graph and box-and-dot plot represent centers and involve spreads. The box-and-dot plot in the middle of Fig. 12.1 simply represents some characteristics of the two samples, while the bullet graph at the right represents some implications that the two samples have about the populations they were selected from. Note also that the scale of the bullet graph is different from the scale of the stem-and-leaf plot and the box-and-dot plot. Even the longest (99% confidence) error bars in the bullet graph are actually substantially shorter than the midspreads indicated in the box-and-dot plot. The scale is enlarged for the bullet graph so that the lengths of the error bars can be seen clearly and compared. If the bullet graph were drawn at the same scale as the box-and-dot plot, the error bars would be so short that they would not be easy to see.

Comparing the two periods on the basis of the error bars at the right of Fig. 12.1 yields the same results that the previous comparisons did with regard to level. Classic period house floors were larger on average than Formative period house floors. This graph, however, also helps us to answer the question about how likely it is that the differences between samples are nothing more than the random variation from one sample to the next that has to be expected even with no real difference between the populations from which the samples are selected.

We estimate that the mean house floor area during the Formative period was $23.8\,m^2 \pm 0.8\,m^2$ at the 80% confidence level. That is, it is not very likely that our Formative sample came from a population with a mean less than $23.0\,m^2$ or greater than $24.6\,m^2$. Our estimate for Classic period house floors is a mean of $26.3\,m^2$. This is substantially outside the 80% confidence level error range for the Formative period. Thus there is less than a 20% chance that the Formative period sample came from a population with a mean as large as $26.3\,m^2$. The Classic period mean is also well outside the 95% confidence level error range for the Formative and even outside the 99% confidence level error range for the Formative. The error range for the Formative at the 99% confidence level reaches only to $25.4\,m^2$, still below the $26.3\,m^2$ mean for the Classic. Thus there is less than a 1% chance that the Formative period sample came from a population with a mean of $26.3\,m^2$. The probability, then, is less than 1% that we would get a sample like our Formative period one from a population like the Classic period population seems to be.

We would arrive at the same conclusion if we made the comparison in the reverse direction by considering how likely it is that a sample like the Classic period one could be selected from a population of house floors like the Formative population seems to have been. The estimated mean for the Formative period falls well outside not only the 80% confidence level error range for the Classic period but also the 95% confidence level error range and the 99% confidence level error range as well.

Extending these one-way comparisons to a simultaneous two-way comparison graphically is only approximate. That is, we have seen that there is less than a 1% chance either of getting the Classic period sample from a population with a mean as low as the Formative period sample's or of getting the Formative period sample from a population with a mean as high as the Classic period sample's. There remains the question of how likely it is that we could get these two samples from the same population (that is, from two populations with the same mean). It is this question that requires a two-way comparison that depends simultaneously on the standard errors of both samples. In order for us to say that the probability of getting the two samples from a single population is less than 1%, the mean of each must lie not just *at*, but *beyond* the 99% confidence level error range for the other. Just how far beyond depends on how unequal the standard errors of the two samples are. If the mean of each sample lies beyond the 99% confidence level error range for the other, however, it is very unlikely that the two samples came from populations with the same mean. The more relevant way to state the implication of this conclusion, of course, is that it is very unlikely that Formative and Classic period house floors had the same mean area. And this, finally, is the conclusion illustrated by the bullet graph in Fig. 12.1, since the mean of each sample lies well beyond the 99% confidence level error range for the other.

Thus the bullet graph in Fig. 12.1 reveals quickly to the eye that Formative period house floors are, on average, some $2.5\,m^2$ smaller than Classic period house floors and that this difference is very unlikely to be the result of the vagaries of sampling. That is, the graph tells us that the sizes and characteristics of the two samples upon which it is based are such as to give us considerable confidence in saying that there was a change in house floor size from Formative to Classic period.

CONFIDENCE, SIGNIFICANCE, AND STRENGTH

It would be more traditional statistical phrasing to say that the difference between Formative and Classic period house floor sizes is very significant. The statistical concept of significance is the mirror image of the concept of confidence as we have been using it. *Confidence* refers to the probability that the results we are stating *are not* attributable just to the vagaries of sampling. *Significance*, on the other hand, refers to the same concept from the opposite perspective – the probability that the results we are stating *are* attributable just to the vagaries of sampling. It is the same thing to say that we have over 99% confidence that there was a difference between house floor areas in the Formative and Classic periods, or to say that there is less than a 1% chance that the difference between our two samples is due to nothing more than the vagaries of sampling. This second way of saying it involves the *significance probability*, which is below 1%. The sum of the probability that corresponds to the level of confidence and the probability that corresponds to the level of significance is always 100%. Positive results in statistics, then, correspond to high confidence probabilities and, at the same time, to low significance probabilities. We are, to repeat, very confident that house floor areas are different in the two periods, which is the same as saying that the difference in house floor area between the two periods is very significant. Very high confidence corresponds to a very high confidence probability; very high significance corresponds to a very low significance probability, since this is the probability that the results we are interested in are nothing more than the vagaries of sampling.

Both confidence and significance are concepts with quite clear and precise meanings in statistics (even if statisticians approach their definitions in many different ways). The notion of confidence in statistics corresponds pretty well to the colloquial use of the word "confidence." In common speech we say we are "confident" about something when we really do not think we are wrong. Paradoxically, the very act of saying that we are confident recognizes the possibility that we might be wrong at the same time that it classifies that possibility as a remote one. (If we really have no doubt at all about a fact, we usually just state it without even bothering to mention that we are quite confident of it.) The colloquial use of "significance," however, is rather different from its statistical use, and it is important not to confuse the two. We are likely to find something "significant," in colloquial speech, if it is important or meaningful. In statistics, however, "significant," like "confident," refers directly to the possibility that the conclusions we are stating are wrong – that is, the possibility that they represent nothing more than the normal variation to be expected in the random sampling process (that is, the vagaries of sampling).

The conclusion we arrived at in this example (that Classic houses were larger than Formative ones) may or may not be meaningful or important, but it *is* very significant. Whether it is meaningful or important is a substantive issue involved with what our interpretation of the result might be. The issue of meaningfulness or importance is an entirely separate one from that of confidence or significance. Staying purely in the realm of statistics, the closest we come to the issue of meaningfulness or importance is in the statistical concept of *strength*. In the comparison we have just made, the notion of strength is quite simple. The strength of the difference in house floor area between Formative and Classic is simply the magnitude of the difference, $2.5\,m^2$ – the amount by which Formative period house floors appear to differ in area from Classic period ones on average.

We are highly confident in identifying this difference; we know that it is very significant – both statements meaning only that the difference we observe in our samples is not at all likely to be just the result of the vagaries of sampling. It is extremely likely that mean house floor size really was greater in the Classic than in the Formative. Whether this result is meaningful or important, however, has to do with why we are interested in this information in the first place. Perhaps we suspect a shift from nuclear family structure in the Formative to extended family structure in the Classic, and we reason that one way this might be evidenced in the archaeological record is in an increase in mean house floor area. We have found a very significant increase in mean house floor area, but it provides little support for our idea because the increase is too small ($2.5\,m^2$) to be seen as an indicator of the need to provide more house space for substantially larger families. Both Formative and Classic period houses are, in general terms, relatively small even for nuclear family groups, and a change of only $2.5\,m^2$ is difficult to relate convincingly to a shift from households of perhaps four or five people to much larger households. Thus the result of our example investigation, while *highly significant*, was *not strong enough* to be important or meaningful, at least in this hypothetical interpretive context.

COMPARISON BY t TEST

The comparison between two samples that we have just made on the basis of bullet graphs of error ranges can also be approached as a significance test problem. This approach is entirely compatible with the comparison as we have already made it – it simply provides a different, and perhaps complementary, perspective on the situation. The t test also makes a comparison that is simultaneously two way in the sense referred to above. The t test enables us to pool all the information from both samples into a single statement of the probability that both could be selected from the same population. Since in such situations we are likely to know that the two samples were, in fact, selected from different populations, this statement is really shorthand for saying "the probability that the two samples were selected from two populations with the same mean."

The two-sample t test evaluates the difference in means between the two samples in light of the pooled standard deviations from both samples. It is as if we calculated error ranges for the two based on the standard deviations from both together so that no matter whether we compared the first to the second or the second to the first, the results would be the same. The equation for accomplishing this at first seems formidable, but evaluating it is really quite a simple process of plugging in familiar values. First, the pooled standard deviation for the two samples is given by the expression

$$s_P = \sqrt{\frac{(n_1 - 1)s_1^2 + (n_2 - 1)s_2^2}{n_1 + n_2 - 2}}$$

where s_P = the pooled standard deviation for the two samples, n_1 = the number of elements in the first sample, n_2 = the number of elements in the second sample, s_1 = the standard deviation in the first sample, and s_2 = the standard deviation in the second sample.

Calculating this quantity for the Formative and Classic period house floor area samples used in the example above produces

$$\begin{aligned} s_P &= \sqrt{\frac{(32 - 1)(3.4)^2 + (52 - 1)(4.5)^2}{32 + 52 - 2}} \\ &= \sqrt{\frac{(358.36 + 1032.75)}{82}} \\ &\quad \sqrt{16.9648} \\ &= 4.12\,\mathrm{m}^2 \end{aligned}$$

This pooled standard deviation for the two samples falls between the standard deviation of 3.4 m² for the Formative sample and the standard deviation of 4.5 m² for the Classic sample – which makes intuitive good sense. The pooled standard deviation is then the basis for a pooled standard error SE_P:

$$SE_P = s_P \sqrt{\frac{1}{n_1} + \frac{1}{n_2}}$$

For the Formative and Classic house floor example,

$$SE_P = 4.12\sqrt{\frac{1}{32} + \frac{1}{52}}$$
$$= 4.12\sqrt{0.0505}$$
$$= 0.93\,\mathrm{m}^2$$

Knowing the pooled standard error enables us to say how many pooled standard errors the difference between sample means represents:

$$t = \frac{\overline{X}_1 - \overline{X}_2}{SE_P}$$

where $\overline{X}_1 =$ the mean in the first sample; and $\overline{X}_2 =$ the mean in the second sample.
    For the house floor area example,

$$t = \frac{23.8 - 26.3}{0.93}$$
$$= -2.69$$

The observed difference in house floor area between the two samples, then, is 2.69 pooled standard errors. We know already that such a large number of standard errors is associated with high statistical confidence, and thus with the low probability values that mean great significance as well. To be more specific, this t value can be looked up in Table 9.1. The number of degrees of freedom is $n_1 + n_2 - 2$, or in this example $32 + 52 - 2 = 82$. Thus we use the row for $60\,d.f.$, which is the closest row to 82. Ignoring the sign for the moment, we look for 2.69 in this row. It would fall between the columns for 1% and 0.5% significance. Thus the probability that the difference we observe between the two samples is just due to the vagaries of sampling is less than 1% and greater than 0.5%. We could also say that the probability of selecting two samples that differ as much as these do from populations with the same mean value is less than 1%. Yet another way to express the same thought is that we are more than 99% confident that average house floor areas differed between the Formative and Classic periods. This is, of course, entirely consistent with the conclusion that was already apparent in Fig. 12.1 and that we have discussed earlier.
    The sign of the t value arrived at indicates the direction of the difference. If the second population has a lower mean than the first, t will be positive. If the second population has a higher mean than the first, t will be negative. The strength of the difference is still indicated simply by the difference in means between the two samples, as it has been all along: $2.5\,\mathrm{m}^2$.

Be Careful How You Say It

When presenting the result of a significance test, it is always necessary to say just what significance test was used, and to provide the resulting statistic, and the associated probability. For the example in the text, we might say, "The 2.5 m^2 difference in mean house floor area between the Formative and Classic periods is very significant ($t = -2.69$, $.01 > p > .005$)." This one sentence really says everything that needs to be said. No further explanation would be necessary if we were writing for a professional audience whom we can assume to be familiar with basic statistical principles and practice. The "statistic" in this case is t, and providing its value makes it clear that significance was evaluated with a t test, which is quite a standard technique that does not need to be explained anew each time it is used. The probability that the observed difference between the two samples was just a consequence of the vagaries of sampling is the *significance* or the *associated probability*. Ordinarily p stands for this probability, so in this case we have provided the information that the significance is less than 1%. This means the same thing as saying that our confidence in reporting a difference between the two periods is greater than 99%.

If, instead of performing a t test, we simply used the bullet graph to compare estimates of the mean and their error ranges, as in Fig. 12.1, we might say "As Fig. 12.1 shows, we can have greater than 99% confidence that mean house floor area changed between Formative and Classic periods." The notion of estimates and their error ranges for different confidence levels is also a very standard one which we do not need to explain every time we use it. Bullet graphs, however, are less common than, say, box-and-dot plots, so we cannot assume that everyone will automatically understand the specific confidence levels of the different widths of the error bars. A key indicating what the confidence levels are, as in Fig. 12.1, is necessary.

In yet another approach, perhaps the most direct of all, we could simply focus on the estimated difference in means and say, "we are 95% confident that house floor area increased by 2.5 m^2 ± 1.9 m^2 from the Formative to the Classic, but this change is not strong enough to connect convincingly to much increase in family size."

In an instance like the example in the text, a bullet graph and a t test are alternative approaches. Using and presenting both in a report qualifies as statistical overkill. Pick the one approach that makes the simplest, clearest, most relevant statement of what needs to be said in the context in which you are writing; use it; and go on. Presentation of statistical results should support the argument you are making, not interrupt it. The simplest, most straightforward presentation that provides complete information is the best.

The pooled standard error from the t test also provides us with an even more direct way to go at the fundamental issue of both the strength and the significance of the difference in means. It enables us to make an estimate of the difference between the means of the two populations involved and to put an error range with that estimate. The best estimate of the difference between the means of the two populations is simply the difference between the means of the two samples. The pooled standard error from the t test is $0.93\,m^2$, and, as usual, this would be an error range for about 66% confidence. With the help of the t table, we can convert this into an error range for 95% confidence in the usual way. The value of t from the table for 82 $d.f.$ and 95% confidence is 2.000, so this is the number of standard errors needed for a 95% confidence error range. Thus, $(2.00)(0.93) = 1.86$, and we can be 95% confident that the difference in means between Formative and Classic period house floor areas is $2.5\,m^2 \pm 1.9\,m^2$, that is between $0.6\,m^2$ and $4.4\,m^2$ larger in the Classic period. This estimate gives us perhaps the most useful of all the ways of presenting the results of all these analyses: we are 95% confident that house area increased by $2.5\,m^2 \pm 1.9\,m^2$ from the Formative to the Classic – a change, but not a large enough change to connect convincingly to much increase in family size.

THE ONE-SAMPLE t TEST

Occasionally we are interested in comparing a sample not to another sample but to some particular theoretical expectation. For example, we might be interested in investigating whether a particular prehistoric group practiced female infanticide. One line of evidence we might pursue would be sex ratios in burials. Suppose we had a sample of 46 burials, which we were willing to take as a random sample of this prehistoric population except for infants intentionally killed, whose bodies we think were disposed of in some other way. On theoretical grounds we would expect this sample of burials to be 50% males and 50% females, unless sex ratios were altered by some practice such as female infanticide. (Actually there might be reason to expect very slightly different proportions from 50:50 on theoretical grounds, but that does not really affect what concerns us here.) After careful study of the skeletal remains, we determine that 21 of the 46 burials were females and 25 were males. The proportions are thus 45.7% female and 54.3% male. This lower proportion of females in our sample might make us think that more females were killed in infancy than males, but we wonder how likely it is that we could select a random sample of 46 with these proportions from a population with an even sex ratio. We could calculate the error ranges for various levels of confidence, as we did in Chapter 11. For a proportion of 45.7% females, in a sample of 46, the standard error would be

$$SE = \frac{\sqrt{pq}}{\sqrt{n}} = \frac{\sqrt{(0.457)(0.543)}}{\sqrt{46}} = \frac{0.498}{6.782} = 0.073$$

For, say, an 80% level of confidence, we would look up the value of t for 80% confidence and 45 degrees of freedom. We would multiply the standard error by this t value: $(0.073)(1.303) = 0.095$. We would thus be 80% confident that our sample of 46 burials was drawn from a population with 45.7% ± 9.5% females (or between 36.2% and 55.2% females). The theoretical expectation of 50% females is well within this 80% confidence level error range, so there is something over a 20% chance that the divergence from even sex ratios that we observe in our sample is only the result of sampling vagaries.

To be more precise about it with a one-sample t test, we would simply use the standard error and the t table in a slightly different way. The observed proportion of females in our sample is 45.7%, or 4.3% different from the expected 50:50 ratio. This difference of 0.043 (that is, 4.3%) represents 0.589 standard errors since $0.043/0.073 = 0.589$. Looking for this number of standard errors on the row of the t table corresponding to 40 degrees of freedom (the closest we can get to 45 degrees of freedom) would put us slightly to the left of the first column in the body of the table (the one that corresponds to 50% significance). There is something over a 50% chance, then, of getting a random sample of 46 with as uneven a sex ratio as this from a population with an even sex ratio. This means that it is uncomfortably likely that the uneven sex ratios we observe in our sample are nothing more than the vagaries of sampling. We might also say, "The difference we observe between our sample and the expected even sex ratio has extremely little significance ($t = .589$, $p > .5$)." These results would not provide much support for the idea of female infanticide. At the same time they would not provide much support to argue that female infanticide was *not* practiced since there is also an uncomfortably large chance that this sample could have come from a population with an uneven sex ratio. In short, given the proportions observed, this sample is simply not large enough to enable us to say with much confidence whether the population it came from had an even sex ratio or not.

THE NULL HYPOTHESIS

Significance tests are often approached by practitioners of many disciplines as a question of testing hypotheses. In this approach, first a *null hypothesis* is framed. In the example of Formative and Classic house floor areas, the null hypothesis would postulate that the observed difference between the two samples was a consequence of the vagaries of sampling. An arbitrary significance level would be chosen for rejecting this hypothesis. (The level chosen is almost always 5%, for no particularly good reason.) And then the t test would be performed. The result ($t = -2.69, 0.01 > p > 0.005$) is a significance level that exceeds the usual 5% rejection level. (That is, the probability that the difference is just due to the vagaries of sampling is even less than the chosen 5% threshold.) Thus the null hypothesis (that the difference is just random sampling variation) is rejected, and the two populations are taken to have different mean areas.

The effect of framing significance tests in this way is to provide a clear "yes" or "no" answer to the question of whether the observation tested really characterizes the populations involved rather than just being the result of sampling vagaries. The problem is that statistics never do really give us a "yes" or "no" answer to this question. Significance tests may tell us that the probability that the observation is just the result of sampling vagaries is very high or moderate or very low. But as long as we are making inferences from samples we are never absolutely certain about the populations the samples represent. Significance is simply not a condition that either exists or does not exist. Statistical results are either more significant or less significant. We have either greater or lesser confidence in our conclusions about populations, but we *never have absolute certainty*. To force ourselves either to reject or accept a null hypothesis is to oversimplify a more complicated situation to a "yes" or "no" answer. (Actually many statistics books make the labored distinction that one does not accept the null hypothesis but rather "fail to reject" it. In practice, analysts often treat a null hypothesis they have been unable to reject as a proven truth – more on this subject later.)

This practice, of forcing statistical results like "maybe" and "probably" to become "no," and "highly likely" to become "yes," has its clearest justification in areas like quality control where an unequivocal "yes" or "no" decision must be made on the basis of significance tests. If a complex machine turns out some product, a quality control engineer may test a sample of the output to determine whether the machine needs to be adjusted. On the basis of the sample results, the engineer must decide either to let the machine run (and risk turning out many defective products if he or she is wrong) or stop the machine for adjustment (and risk wasting much time and money if he or she is wrong). In such a case, statistical results like "the machine is probably turning out defective products" must be converted into a "yes" or "no" answer to the question of stopping the machine. Fortunately, research archaeologists are rarely in such a position. We can usually (and more informatively) say things like "possibly," "probably," "very likely," and "with great probability."

Finally, following the traditional 5% significance rule for rejecting the null hypothesis leaves us failing to reject the null hypothesis when the probability that our results are just the vagaries of sampling is only 6%. If, in the house floor example, the t value had been lower, and the associated probability had been 6%, it would have been quite reasonable for us to say, "We have fairly high confidence that mean house floor area was greater in the Classic period than in the Formative." If we had approached the problem as one of attempting to reject a null hypothesis, however, with a 5% rejection level, we would have been forced to say instead, "We have failed to reject the hypothesis that house floor areas in the Formative and Classic are the same." As a consequence we would probably have proceeded as if there were no difference in house floor area between the two periods when our own statistical results had just told us that there was a 94% probability that there *was* such a difference.

In some disciplines, almost but not quite rejecting the null hypothesis at the sacred 5% level is dealt with by simply returning to the lab or wherever and studying a larger sample. Other things being equal, larger samples produce higher confidence levels, and higher confidence levels equate to lower significance probabilities.

Table 12.3. Summary of Contrasting Approaches to Significance Testing in the Context of the House Floor Area Example

Significance testing as an effort to reject the null hypothesis (not recommended here).	Significance testing as an effort to evaluate the probability that our results are just the vagaries of sampling (the approach followed in this book).
The questions asked:	
The difference observed between the Formative and Classic house floor samples is nothing more than the vagaries of sampling. True or false?	How likely is it that the difference observed between Formative and Classic house floor samples is nothing more than the vagaries of sampling?

Example answers for different possible significance levels:

$p = .80$	True.	Extremely likely.
$p = .50$	True.	Very likely.
$p = .20$	True.	Fairly likely.
$p = .10$	True.	Not very likely.
$p = .06$	True.	Fairly unlikely.
$p = .05$	False.	Fairly unlikely.
$p = .01$	False.	Very unlikely.
$p = .001$	False.	Extremely unlikely.

Almost but not quite rejecting the null hypothesis, then, can translate into "There is probably a difference, but the sample was not large enough to make us as confident as we would like about it." In archaeology, however, it is often difficult or impossible to simply go get a larger sample, so we need to get all the information we can from the samples we have. For this reason, in this book we will approach significance testing not as an effort to reject a null hypothesis but instead as an effort to say just how likely it is that the result we observe is attributable entirely to the vagaries of sampling.

Table 12.3 summarizes the differences between a null hypothesis testing approach to significance testing and the more scalar approach advocated in this book. The approach followed here can, of course, be thought of as testing the null hypothesis but not forcing the results into a "yes" or "no" decision about it. It we are willing to take a more scalar approach, though, there is no advantage in plunging into the confusion of null hypothesis formulation, rejection, and failure to reject. In particular, Table 12.3 emphasizes how potentially misleading is the answer "true" when applied to a full range of probabilities concerning the null hypothesis that can more accurately be described as ranging from "extremely likely" to "fairly unlikely."

Pregnancy tests have only two possible results: pregnant and not pregnant. Significance tests are simply not like that; their results run along a continuous scale of variation from very high significance to very low significance. While some users of statistics (poker players, for example) find themselves having to answer "yes" or "no" questions on the basis of the probabilities given by significance tests, archaeologists can count themselves lucky that they are not often in such a situation. We are almost always able to say that results provide very strong support for our ideas, or moderately strong support, or some support, or very little support. Forcing significance tests simply to reject or fail to reject a null hypothesis, then, is usually unnecessary and unhelpful in archaeology and may do outright damage by being misleading as well. In this book we will never characterize results as simply "significant" or "not significant" but rather as more or less significant with descriptive terms akin to those in Table 12.3. Some statistics books classify this procedure as a cardinal sin. Others find it the only sensible thing to do. The fact is that neither approach is truth revealed directly by God. Archaeologists must decide which approach best suits their needs by understanding the underlying principles, not by judging which statistical expert seems most Godlike in revealing his or her "truth."

STATISTICAL RESULTS AND INTERPRETATIONS

It is easy to accidentally extend levels of confidence or significance probabilities beyond the realm to which they properly apply. Either one is a statistical result that takes on real meaning or importance for us only when interpreted. In the example of Formative and Classic period house floors used throughout this chapter, our interest, as suggested earlier, may be investigating a possible shift from nuclear families in the Formative to extended families in the Classic. Given the samples we have in this example, we find that houses in the Classic were larger, on average, than houses in the Formative. We also find that this difference is very significant (or that we have quite high confidence that it is not just the result of sampling vagaries, which means the same thing). This does not, however, automatically mean that we have quite high confidence that nuclear family organization in the Formative changed to extended family organization in the Classic. The former is a statistical result; the latter is an interpretation. How confident we are in this latter interpretation depends on a number of things in addition to the statistical result. For one thing, already mentioned early in this chapter, despite the high significance of the size difference between Formative and Classic house floors, the strength or size of the difference $(2.5\,m^2 \pm 1.9\,m^2)$ is not very much – at least not compared to what we would expect from such a change in family structure. There might be several other completely different kinds of evidence we could bring to bear as well. It would only be after weighing all the relevant evidence (among which our house floor area statistical results would be only one item) that we would be prepared to assess how much confidence we have in the suggested interpretation. We would not really be in position to put a number on our confidence in the interpretation about family structure, because it is an interpretation of the evidence, not a statistical result. We would presumably need to weigh the family structure interpretation against other possible interpretations of the evidence. While statistical evaluation of the various lines of evidence is extremely helpful in this process, its help comes from evaluating the confidence we should place in certain patterns observable in the measurements we make on the

samples we have, not from placing probabilities directly on the interpretations themselves. These interpretations are connected sometimes by a very long chain of more
or less convincing logical links and assumptions to the statistical results obtained by
analyzing our samples.

ASSUMPTIONS AND ROBUST METHODS

The two-sample t test assumes that both samples have approximately normal shapes
and roughly similar spreads. If the samples are large (larger than 30 or 40 elements)
violations of the first assumption can be tolerated because the t test is fairly robust.
As long as examination of a box-and-dot plot reveals that the midspread of one
sample is no more than twice the midspread of the other, the second assumption can
be considered met. If the spreads of the two samples are more different than this,
then that fact alone suggests that the populations they came from are different, and
that, after all, is what the two-sample t test is trying to evaluate.

If the samples to be compared contain outliers, the two-sample t test may be
very misleading, based as it is on means and standard deviations, which will be
strongly affected by the outliers, as discussed in Chapters 2 and 3. In such a case
an appropriate approach is to base the t test on the trimmed means and trimmed
standard deviations, also discussed in Chapters 2 and 3. The calculations for the t
test in this case are exactly as they are for the regular t test except that the trimmed
sample sizes, the trimmed means, and the trimmed standard deviations are used
in place of the regular sample sizes, the regular means, and the regular standard
deviations.

If the samples to be compared are small and have badly asymmetrical shapes,
this can be corrected with transformations, as discussed in Chapter 5, before performing the t test. The data for both samples are simply transformed and the t test is
performed exactly as described above on the transformed batches of numbers. It is,
of course, necessary to perform the same transformation on both samples, and this
may require a compromise decision about which transformation produces the most
symmetrical shape simultaneously for both samples.

For samples with asymmetrical shapes, of course, estimating the median in the
population instead of the mean makes good sense, and putting error ranges with estimates of the median with the bootstrap was discussed in Chapter 10. The median and
those error ranges could be used instead of the mean as the basis for a bullet graph
like the one in Fig. 12.1 for a graphical comparison. Yet another kind of graph adds
error ranges to the box-and-dot plot, which we already examined for comparing the
medians of batches in Chapter 4. The notched box-and-dot plot in Fig. 12.2 compares the batches of Early and Late Classic site areas from the bootstrap example in
Chapter 10. The notches in each box have their points at the estimated population
median and their ends at the top and bottom of its error range.

Figure 12.2 A notched box-and-dot plot comparing Early and Late Classic site areas.

The error ranges represented in notched box plots are usually not just the 95 or 99% confidence error ranges that we have used for bullet graphs. Often they represent a specially contrived error range a bit like the pooled standard error from the *t* test. If the upper limit of one error range just reaches the level of the lower limit of the other error range, then the probability that the two samples came from populations with the same median is roughly 5%. If the error ranges for the two batches represented by the notches do not overlap, then we can be over 95% confident that the two samples came from populations with different medians. It is this kind of error range that appears in the notched box-and-dot plots in Fig. 12.2, and so the comparison works a bit differently than the comparison in the bullet graph in Fig. 12.1, which has been drawn to represent straightforward error ranges for particular confidence levels. Both bullet graphs and notched box-and-dot plots indicate our confidence that there is a difference between the two populations only approximately, but they do it differently. For bullet graphs, the focus is not on whether the ends of the 95% confidence error ranges overlap, but on whether the estimated mean for one population falls beyond the error range for the other, and vice versa.

In Fig. 12.2, the notches overlap quite substantially, and this lets us know that we are quite substantially less than 95% confident that median site area changed from Early to Late Classic. This result would not encourage us to spend much time pondering what caused the slight increase in median site area seen in the Late Classic sample, because, with such low confidence that there was any difference at all between these two populations, we might very well be seeking a reason for something that had not even occurred.

PRACTICE

You have just completed extensive excavations at the Ollantaytambo site. You select a random sample of 36 obsidian artifacts from those recovered at the site for trace element analysis in an effort to determine the source(s) of the raw material. You realize that you really should investigate a whole suite of elements, but the geochemist you collaborate with gives you only the data for the element zirconium before he sets off down the Urubamba River in a dugout canoe, taking with him the remainder of the funds you have budgeted for raw material sourcing. There are two visually different kinds of obsidian in the sample – an opaque black obsidian and a streaky gray obsidian – and you know that such visual distinctions sometimes correspond to different sources. The data on amounts of zirconium and color for your sample of 36 artifacts are given in Table 12.4.

1. Begin to explore this sample batch of zirconium measurements with a back-to-back stem-and-leaf diagram to compare the black and gray obsidian. What does this suggest about the source(s) from which black and gray obsidian came?
2. Estimate the mean measurement for zirconium for gray obsidian and for black obsidian in the populations from which these samples came. Find error ranges for these estimates at 80%, 95% and 99% confidence levels, and construct a bullet graph to compare black and gray obsidian. How likely does this graph make it seem that gray and black obsidian came from a single source?

Table 12.4. Zirconium Content for a Sample of Gray and Black Obsidian Artifacts from Ollantaytambo

Zirconium content (ppm)	Color	Zirconium content (ppm)	Color
137.6	Black	136.2	Gray
135.3	Gray	139.7	Gray
137.3	Black	139.1	Black
137.1	Gray	139.2	Gray
138.9	Gray	132.6	Gray
138.5	Gray	134.3	Gray
137.0	Gray	138.6	Gray
138.2	Black	138.6	Black
138.4	Black	139.0	Black
135.8	Gray	131.5	Gray
137.4	Black	142.5	Black
140.9	Black	137.4	Gray
136.4	Black	141.7	Black
138.8	Black	136.0	Gray
136.8	Gray	136.9	Black
136.3	Gray	135.0	Gray
135.1	Black	140.3	Black
132.9	Gray	135.7	Black

3. What, exactly, have you calculated the probabilities of in Question 2? What, exactly, are the populations you have characterized? What are the logical links necessary to use this evidence in support of the conclusion you want to make about obsidian sources?

4. Approach Question 2 using a *t* test. How strong and significant is the difference in zirconium content between black and gray obsidian? What is your estimate of that difference with an error range at the 95% confidence level? State the conclusions derived from your *t* test in a single clear sentence as if you were reporting it in a paper.

Chapter 13
Comparing Means of More than Two Samples

In Chapter 12 we took two approaches to comparing the means of two samples. The first approach involved using each sample separately to estimate the mean of the population that the sample came from. We then attached error ranges for several confidence levels to these estimates and drew a picture of the whole thing with a bullet graph (Fig. 12.1). This approach is easily extended to the comparison of any number of samples. In this chapter we will use another fictitious example consisting of 127 Archaic projectile points from the Cottonwood River valley. After considering possible sources of bias we decide to work with these as a random sample from the large and vaguely defined population of Archaic projectile points from the Cottonwood River valley.

We are interested in whether, during the Archaic period, there was much change in hunting of large and small animals in the Cottonwood River valley. We reason that large projectile points are more involved in hunting large animals and small projectile points are more involved in hunting small animals. We can divide the 127 projectile points into three groups: Early, Middle, and Late Archaic, and we decide to compare the weights of projectile points in these three periods. One way to organize these data for this sample is shown in Table 13.1. Here two observations are recorded for each of the 127 projectile points: the weight (in grams) and the period (Early, Middle, or Late Archaic). Our two variables, weight and period, are of different kinds. Weight, of course, is a measurement, and period is a set of three categories.

R.D. Drennan, *Statistics for Archaeologists*, Interdisciplinary Contributions
to Archaeology, DOI 10.1007/978-1-4419-0413-3_13,
© Springer Science+Business Media, LLC 2004, 2009

Table 13.1. Data on Weight and Period for a Sample of Archaic Period Projectile Points from the Cottonwood River Valley

Weight (g)	Archaic Subperiod	Weight (g)	Archaic Subperiod	Weight (g)	Archaic Subperiod	Weight (g)	Archaic Subperiod
54	Early	68	Early	63	Middle	69	Middle
39	Early	68	Early	52	Middle	80	Middle
49	Early	85	Early	44	Middle	78	Middle
65	Early	49	Early	73	Middle	69	Middle
54	Early	21	Early	70	Middle	34	Late
83	Early	24	Early	56	Middle	39	Late
75	Early	50	Early	46	Middle	40	Late
45	Early	52	Early	61	Middle	45	Late
68	Early	62	Early	49	Middle	37	Late
47	Early	44	Early	51	Middle	32	Late
57	Early	61	Early	61	Middle	31	Late
19	Early	30	Early	70	Middle	60	Late
47	Early	52	Early	51	Middle	58	Late
58	Early	56	Early	42	Middle	45	Late
76	Early	63	Early	73	Middle	50	Late
50	Early	53	Early	51	Middle	40	Late
67	Early	79	Early	74	Middle	41	Late
52	Early	50	Early	40	Middle	38	Late
40	Early	54	Early	67	Middle	59	Late
58	Early	51	Early	51	Middle	37	Late
42	Early	59	Early	59	Middle	28	Late
43	Early	60	Early	68	Middle	37	Late
58	Early	48	Early	63	Middle	31	Late
28	Early	40	Early	64	Middle	40	Late
59	Early	50	Early	78	Middle	34	Late
43	Early	69	Early	62	Middle	37	Late
45	Early	71	Middle	78	Middle	44	Late
60	Early	64	Middle	57	Middle	47	Late
27	Early	59	Middle	59	Middle	54	Late
64	Early	65	Middle	31	Middle	36	Late
73	Early	54	Middle	69	Middle	48	Late
70	Early	65	Middle	32	Middle		

COMPARISON WITH ESTIMATED MEANS AND ERROR RANGES

We can use the three period categories to separate the sample of 127 projectile points into three samples – one consisting of the 58 Early Archaic points, one of the 42 Middle Archaic points, and one of the 27 Late Archaic points. If we were willing to treat the 127 projectile points as a random sample from the Archaic projectile points of the Cottonwood River valley, then we can be equally willing to treat

Table 13.2. Comparison of Weights of Projectile Points for
Archaic Subperiods

	Early	Middle	Late	All Archaic
$n =$	58	42	27	127
$\overline{X} =$	53.67 g	60.45 g	41.56 g	53.34 g
$s =$	14.67 g	12.15 g	8.76 g	14.42 g
$SE =$	1.93 g	1.88 g	1.69 g	1.28 g
$s^2 =$	215.21	147.62	76.74	207.94

the 58 Early Archaic points as a random sample of the Early Archaic points of the Cottonwood River valley, the 42 Middle Archaic points as a random sample of the Middle Archaic projectile points of the Cottonwood River valley, and the 27 Late Archaic points as a random sample of the Late Archaic points of the Cottonwood River valley. If we do this, then we have reorganized a single batch of numbers into three batches of numbers that can be compared just as we compared the two batches of numbers in Chapter 12.

Table 13.2 provides numerical indexes (sample size, mean, standard deviation, standard error, and variance) for each of these three smaller samples. Using the standard errors and Table 9.1 we can provide estimated mean weights for each of the three populations these samples came from and present the whole comparison graphically as in Fig. 13.1. The Early Archaic and Middle Archaic samples are large enough that we can count on a special batch with a normal shape. The Late Archaic sample is a bit small for us to count on a normal shape for the special batch, so we look at the stem-and-leaf plot for Late Archaic in Fig. 13.1 to make sure that the sample itself has a fairly normal shape (which it does). The box-and-dot plot makes clear that Middle Archaic projectile points tend to be the heaviest and Late Archaic projectile points the lightest, with Early Archaic projectile point weights falling in between. The ranges of all three certainly overlap, however, especially those of Early Archaic and Middle Archaic. At the far right in Fig. 13.1, the bullet graph of estimated population means with error bars for confidence levels of 80%, 95%, and 99% makes it clear that the differences between the three samples we have are very highly significant. None of the error ranges for 99% confidence includes the estimated mean of any of the other populations. We are thus more than 99% confident that the differences we observe between samples are not just a consequence of the vagaries of sampling. It is extremely likely instead that such different samples came from parent populations that differed from each other.

Figure 13.1 demonstrates once again that box-and-dot plots and bullet graphs are two different things. The boxes representing the midspreads for the three periods overlap substantially, while the error ranges for 80%, 95%, and 99% confidence do not. Since these two kinds of plots are similar in appearance and since both deal with the spreads of batches in one way or another, it is easy to overlook the fundamental difference between the two. While it is true that the error ranges in the bullet graph in Fig. 13.1 are based, in part, on the spread of each batch, they are not simply a graphical representation of that spread. They rely as well on the sample sizes and are

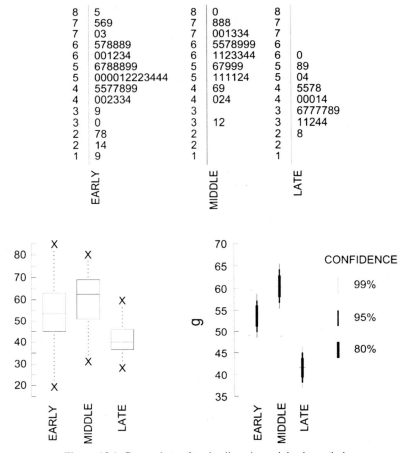

Figure 13.1. Comparison of projectile point weights by period.

thus a picture not of the spreads of the three sample batches but rather of the spreads of the corresponding special batches, as discussed in Chapter 8. As a consequence, the bullet graphs have useful implications concerning the parent populations that the box-and-dot plots do not have.

COMPARISON BY ANALYSIS OF VARIANCE

Estimating means and attaching error ranges to each estimate provides a good way to compare each sample with each other sample, and a bullet graph literally draws the overall picture. In terms of significance, this overall picture is summed up in the question, "How likely is it that we could get three samples with means and standard deviations like these from a single parent population?" Another way to say it would be, "What is the probability that samples as different as these three could be produced from the same population just through the vagaries of sampling?" When we speak of a "single parent population" or the "same population" in these questions, we are speaking metaphorically, since we know the three samples came from three different populations, one of Early Archaic projectile points, one of Middle Archaic projectile points, and one of Late Archaic projectile points. Hypothesizing here that they may have come from the "same population" is simply shorthand for inquiring how likely it is that the three populations these three samples came from had the same mean. Thus, the significance question we are asking amounts to, "How likely is it that Early Archaic, Middle Archaic, and Late Archaic projectile point populations all had the same mean weight, and that our three samples differ just because random samples, even from the same population, do differ from each other?"

We answered such a question with a two-sample t test in Chapter 12, but this test cannot easily be extended to more than two samples. For three or more samples, the technique of choice is *analysis of variance*, often abbreviated *ANOVA*. As the name implies, analysis of variance relies on variance as the key to answering the significance question in this situation. (Remember that the variance of a batch is simply the square of the standard deviation.) The variances (s^2) of all three separate subsamples and of the entire sample of 127 are given in Table 13.2.

Analysis of variance assumes that the samples are drawn from populations with normal shapes. We examine the stem-and-leaf plots for the three separate subsamples in Fig. 13.2, and we see the fundamentally single-peaked and symmetrical shape that we need to see for each of the subsamples. Analysis of variance also assumes that the spreads (specifically the variances) of the populations are approximately equal. The box-and-dot plots in Fig. 13.1 provide an easy way to judge the spreads of the samples, as do the figures for the variances given in Table 13.2.

```
8 | E            8 |            8 |            8 | E
7 | EEE          7 |            7 | L          7 | EEEL
7 | EE           7 | M          7 | LL         7 | EEMLL
6 | EEEEEE       6 | MMM        6 | LL         6 | EEEEEMMMLL
6 | EEEEEE       6 | MMMMMM     6 | LLLL       6 | EEEEEEMMMMMMLLLL
5 | EEEEEEE      5 | MMMMMMM    5 | LLLLL      5 | EEEEEEEMMMMMMMLLLLL
5 | EEEEEEEEEEEE 5 | MMMMMMM    5 | LLLLLLL    5 | EEEEEEEEEEEEMMMMMMMMLLLLLLL
4 | EEEEEEE      4 | MMMMM      4 | LLLLL      4 | EEEEEEEMMMMMLLLLL
4 | EEEEE        4 | MMMMMM     4 | L          4 | EEEEEEMMMMMML
3 | E            3 | MM         3 |            3 | EMM
3 | E            3 | MMM        3 |            3 | EMMM
2 | EE           2 |            2 |            2 | EE
2 | EE           2 | MM         2 |            2 | EEMM
1 | E            1 |            1 |            1 | E
    EARLY            MIDDLE        LATE            ALL ARCHAIC
```

Figure 13.2. Stem-and-leaf plots of projectile point weights by subperiod where all subperiod groups have similar means.

Here the largest variance is almost three times as big as the smallest. Comparing midspreads in the box-and-dot plots yields a similar observation. This difference in spreads is pressing the limits of analysis of variance's ability to withstand violation of its basic assumptions. As long as the largest variance is no more than three times the smallest, though, we are willing to go ahead and perform analysis of variance, especially if the samples involved are not too small.

Figure 13.2 illustrates one possible result of a comparison of weights for the three subsamples of projectile points from different parts of the Archaic period. Note that Fig. 13.2 does not really illustrate the data presented in Table 13.1. Instead, it illustrates one pattern that we *might* have seen. This pattern has been created by maintaining the real shapes of all three subsamples but shifting their centers so that they fall much closer together for purposes of discussion only. The stem-and-leaf plots in Fig. 13.2 are drawn with letters standing for the different subperiods in order to make it possible to see what happens when the three subsamples are combined, as at the extreme right.

When we compare the overall sample of 127 projectile points in Fig. 13.2 to the individual subsamples, we observe several things. First, in this result, all three subsamples look pretty much the same. All three have centers in about the same place. All three have roughly similar spreads. Second, the spread of the overall sample of 127 projectile points is similar to the spreads of the individual subsamples. And third, the center of the overall sample of 127 projectile points is quite similar to the centers of the individual subsamples. Despite some minor differences in shape, all four stem-and-leaf plots are fairly similar. The sharpest difference is that the peak in the stem-and-leaf for the overall sample is considerably higher than the peaks for the individual subsamples. This should not be surprising, since the overall sample has considerably more projectile points, but a spread not really larger than those of the individual subsamples. Consequently they mount up higher at the peak.

A different possible result of such a comparison is illustrated in Fig. 13.3, and this figure *does*, in fact, accurately reflect the data in Table 13.1. Comparing Fig. 13.3 with Fig. 13.2 reveals the nature of the differences. First, the three subsamples no longer look pretty much the same. Their spreads continue to be roughly similar, but their centers are clearly in different places. Second, the spread of the overall sample is larger in Fig. 13.3 than in Fig. 13.2. It is no longer as close to the spreads of the individual subsamples as it was in Fig. 13.2. While the Early Archaic subsample has the largest spread, and this continues to be comparable to the spread in the overall sample, the Middle Archaic and Late Archaic subsamples to have noticeably narrower spreads than the overall batch. And third, the center of the overall sample, while similar to the center in the Early Archaic subsample, is distinctly lower than the center in the Middle Archaic subsample and distinctly higher than the center in the Late Archaic subsample.

In sum, Fig. 13.3 shows that, as the centers of the subsamples vary from each other, greater variation is introduced into the overall sample when the three subsamples are combined. Figure 13.2 illustrates a situation where all three subsamples might well have been selected from populations with the same means. Figure 13.3 illustrates a situation where it is considerably more likely that the three subsamples

```
8 | E             8 | M         8 |            8 | EM
7 | EEE           7 | MMM       7 |            7 | EEEMMM
7 | EE            7 | MMMMMM    7 |            7 | EEMMMMMM
6 | EEEEE         6 | MMMMMMM   6 |            6 | EEEEEEMMMMMM
6 | EEEEEE        6 | MMMMMMM   6 | L          6 | EEEEEEMMMMMMMO
5 | EEEEEE        5 | MMMMM     5 | LL         5 | EEEEEEMMMMMLL
5 | EEEEEEEEEEEE  5 | MMMMMM    5 | LL         5 | EEEEEEEEEEEEEEMMMMMMLL
4 | EEEEEE        4 | MM        4 | LLLL       4 | EEEEEEEMMLLLL
4 | EEEEE         4 | MMM       4 | LLLLL      4 | EEEEEEMMMLLLLL
3 | E             3 |           3 | LLLLLLL    3 | ELLLLLLL
3 | E             3 | MM        3 | LLLLL      3 | EMMLLLLL
2 | EE            2 |           2 | L          2 | EEL
2 | EE            2 |           2 |            2 | EE
1 | E             1 |           1 |            1 | E

     EARLY            MIDDLE        LATE          ALL ARCHAIC
```

Figure 13.3. Stem-and-leaf plots of projectile point weights by subperiod for the data presented in Table 13.1.

were selected from populations with different means. Analysis of variance finds the key to assessing these probabilities in a comparison between the variance observed *between* subsamples on the one hand and the variance observed *within* subsamples on the other. These two variances, *between groups* and *within groups*, are calculated very much like the variances of ordinary batches of measurements.

Recall the equation for variance from Chapter 3:

$$s^2 = \frac{\sum (x - \overline{X})^2}{n - 1}$$

The numerator of this fraction, $\sum (x - \overline{X})^2$, is often referred to as the *sum of squares* since it consists of the sum of the squares of the deviations from the sample mean of all the elements in the sample. The denominator, $n - 1$, is actually the number of *degrees of freedom*, a term we did not use in Chapter 3, but which we have come across since.

To calculate the *between groups* variance needed for analysis of variance we must determine what the relevant sum of squares is and what the relevant number of degrees of freedom is. The between groups sum of squares is

$$SS_B = \sum n_i (\overline{X}_i - \overline{X}.)^2$$

where SS_B = the between groups sum of squares, n_i = the number of elements in the ith group (or subsample), $\overline{X}_i$ = the mean in the ith group (or subsample), and $\overline{X}.$ = the mean of all the groups (taken together).

In our example, there are three groups or subsamples, so i refers, in turn, to each of the three groups, whose numerical indexes are given in Table 13.2. Thus

$n_1 = 58$ (the number of Early Archaic projectile points);
$n_2 = 42$ (the number of Middle Archaic projectile points);
$n_3 = 37$ (the number of Late Archaic projectile points;
$\overline{X}_1 = 53.67\,g$ (the mean Early Archaic projectile point weight);
$\overline{X}_2 = 60.45\,g$ (the mean Middle Archaic projectile point weight);
$\overline{X}_3 = 41.56\,g$ (the mean Late Archaic projectile point weight);
$\overline{X}. = 53.34\,g$ (the grand mean projectile point weight, that is, the mean of the total sample including all periods).

Consequently,

$$SS_B = 58(53.67 - 53.34)^2 + 42(60.45 - 53.34)^2 + 27(41.56 - 53.34)^2$$
$$= 6.32 + 2123.19 + 3746.75$$
$$= 5876.26$$

The relevant number of degrees of freedom for this between groups sum of squares is one less than the number of subsamples. For this example, there are three subsamples, so there are two degrees of freedom. Dividing the between groups sum of squares by the number of degrees of freedom, we get

$$s_B^2 = \frac{SS_B}{d.f.} = \frac{5876.26}{2} = 2938.13$$

This figure is the between groups variance, often referred to as the *between groups mean square*. It is the way we express the spread observed between the means of the different groups for analysis of variance.

Analysis of variance seeks to compare the between groups variance just calculated to the within groups variance. This within groups variance, like the between groups variance, involves dividing a sum of squares by the relevant number of degrees of freedom. It amounts to pooling the separate variances of the subsamples. The within groups sum of squares is obtained simply by multiplying each subsample's variance by one less than the number in the subsample and adding up the results for all the subsamples:

$$SS_W = \sum (n_i - 1)s_i^2$$

where SS_W = the within groups sum of squares, n_i = the number of elements in the ith group (or subsample) as before, and s_i^2 = the variance of the ith group (or subsample).

Finding the variances of the subsamples in Table 13.2 gives us the following values for s_i^2: $s_1^2 = 215.21$, $s_2^2 = 147.62$, and $s_3^2 = 76.74$. Consequently, in our example,

$$SS_W = (58-1)(215.21) + (42-1)(147.62) + (27-1)(76.74)$$
$$= 12266.97 + 6052.42 + 1995.24$$
$$= 20314.63$$

The relevant number of degrees of freedom for this within groups sum of squares is the overall sample size minus the number of subsamples. In our example, the overall sample size is 127, and there are three subsamples, so the within groups number of degrees of freedom is 124. Dividing the within groups sum of squares by the number of degrees of freedom, we get

$$s_W^2 = \frac{SS_W}{d.f.} = \frac{20314.63}{124} = 163.83$$

This figure is the within groups variance, often referred to as the *within groups mean square*. It is the expression of the spread to be observed within the various groups needed for analysis of variance.

Once the between groups variance and the within groups variance are calculated, the analysis of variance is almost complete. It only remains to express these two variances as a ratio:

$$F = \frac{s_B^2}{s_W^2}$$

This F ratio in our example comes to

$$F = \frac{2938.13}{163.83} = 17.93$$

The F ratio can then be looked up in a table providing probabilities associated with the different values of F. The probability associated with an F ratio of 17.93 for 2 degrees of freedom between groups and 124 degrees of freedom within groups is 0.0000001. This means that there is only one chance in ten million of randomly selecting three subsamples with the means and standard deviations that these have from three populations whose means are the same. There is, then, a vanishingly small probability that the differences observed between these three samples are simply a consequence of the vagaries of sampling. Our results are extremely significant. We have extremely high confidence that projectile points from different periods really do have different mean weights.

Table 13.3. Example Computer Output for the Analysis of Variance Example in This Chapter

ANALYSIS OF VARIANCE					
SOURCE	SUM OF SQUARES	DF	MEAN SQUARE	F	PROBABILITY
BETWEEN GROUPS	5880.6	2	2940.30	17.94	0.0000001
WITHIN GROUPS	20321.8	124	163.89		

STRENGTH OF DIFFERENCES

In addition to discussing the significance of the differences we observed in the mean weights of projectile points from different parts of the Archaic, we should discuss the strength of these differences. The strength of the differences amounts to nothing more complicated than the differences of means between the subsamples. Late Archaic projectile points are the lightest, on average, with a mean 12.11 g below that of Early Archaic points, which are, in turn, 6.78 g lighter than Middle Archaic ones. Thus the sharpest contrast is the 18.89 g that separate the mean for Middle Archaic points from the mean for Late Archaic points. These are, of course,

---

Statpacks

Having gone through the lengthy calculations in the text, we must recognize that calculating an analysis of variance by hand is largely an outmoded technique. Although there are computational shortcuts that make it easier, and many statistics books provide detailed instructions for these shortcuts, there is not much reason to perform an analysis of variance now except with a computer. (One reason, though, would be if outliers made you want to perform it with trimmed means and standard deviations.) The ease of performing an analysis of variance with a statpack makes it even more important to understand what one is about, and what the resulting numbers mean. The point of the example is to help make clear how analysis of variance works, more than to provide instructions for how to do it.

The details of performing an analysis of variance vary from one statpack to another. Most will want the data organized as they are in Table 13.1. The Archaic subperiod in that table may be called a *grouping variable* or *independent variable* and the weight may be called a *dependent variable*. The output will likely list the between groups and within groups sums of squares, degrees of freedom, and mean squares. The F ratio will be provided, along with its associated probability. Table 13.3 shows an example of one statpack's output for the analysis of this example. This output is from SYSTAT®, and you will note that the numbers are slightly different from those in the text. This is a consequence of rounding error. Statpacks customarily keep track of figures throughout the calculations to many more decimal places than is possible with an ordinary calculator, and thus they produce results with greater precision, although there is no substantive difference in conclusions. Since statpacks also calculate the associated probability with much more accuracy than even a very long and detailed F table can provide, and since the need to performing an analysis of variance by hand does not often arise, this book does not include an F table.

the differences whose significance we have been evaluating, first by comparing estimated means and error ranges in a bullet graph and later through analysis of variance. Both strength and significance can be seen in Fig. 13.1, and that is a clear advantage of such a presentation. It is easy to identify there exactly which subsamples are heavy and which are light and by approximately how much. In fact, most of what we need to say about these numbers is most easily seen in Fig. 13.1. For most purposes in archaeology, a bullet graph is much simpler and more straightforward than an analysis of variance. If it is important to put a single probability figure on the entire pattern of subsamples, however, analysis of variance is available.

Whether the results of our analysis are meaningful depends on both significance and strength but in different ways. If there is very little significance, then there is little point in discussing the meaning of the differences observed, because there is too high a probability that we would simply be discussing the random differences between three samples selected from indistinguishable populations. If there is moderate to high significance, then the strength or magnitude of the differences is worth discussing, at least tentatively, because it seems likely that there is a "real" difference to be discussed. If the significance level is very high, then it is worth engaging in serious discussion of the strength of the differences. Even though the significance level is extremely high in our example, this does not automatically make the results meaningful. It makes them very likely to be "real," but many "real" things are trivial. Whether this difference means anything depends on the substantive issues that we are investigating. If smaller projectile points were, indeed, used for hunting smaller animals, then our results might be used to support an interpretation that smaller animals were most hunted in the Late Archaic and larger animals were most hunted in the Middle Archaic, with the Early Archaic falling somewhere in between. This, then, would imply a shift from smaller toward larger and then back toward smaller game. Whether the 10–20 g involved in mean weight differences is large enough to be meaningful in this context is a substantive rather than statistical evaluation. And, of course, as always, we would want to look at other completely different lines of evidence relevant to the issue, such as site locations, faunal remains, and many others.

As discussed in Chapter 12, significance and strength are two importantly different concepts. Significance is, in some sense, the more "purely" statistical of the two, while strength usually sets us on the path toward the substantive interpretation of the statistical results. Only when relatively high significance is combined with strong enough results to have substantive meaning do our statistical results have much importance. Highly significant results may have little meaning because they are very weak, and very strong results may have little importance because their significance level is low.

DIFFERENCES BETWEEN POPULATIONS VERSUS RELATIONSHIPS BETWEEN VARIABLES

Analysis of variance can also be thought about from a rather different perspective. Instead of focusing on the differences between several populations in mean values of some measurement, we could focus on the analysis of variance as an investigation of the relationship between two variables. In the example above,the two variables would be projectile point weight and period. In an analysis of variance, conceived in this way, there are always two variables: one of them is a measurement, and the other is a set of categories. It is the categorical variable that provides the basis for the division of the overall sample into subsamples, one corresponding to each category.

The categorical variable is always considered the *independent variable* because we simply take the division of the sample into subsamples based on these categories as a given. The measurement is called the *dependent variable* because we speak as if it were determined, at least in part, by the categories. In the example of Archaic projectile points from the Cottonwood River valley we found that Late Archaic projectile points weighed less, on average, than Early Archaic ones. Thus it seems reasonable to say that projectile point weight *depends on* period to some extent. It is simpler in statistics to speak of the relationship in these terms, although this implies nothing about the direction of causality in the real world. Indeed, it makes little real sense even to talk about period as an independent variable that "causes" projectile points to be larger or smaller. This is simply a convention of statistical language, having little to do with real notions of causality.

It is often useful to think of variable relationships in predictive terms. If the two variables – projectile point weight and period – are related to each other, then knowing the value of one for a particular case would help us to predict the value of the other. If, before looking at a particular projectile point, we wished to predict its weight, the best guess we could make would be the mean of the overall sample. That guess would most often be closest to the real weight of the projectile point in question. Given what we found out in the analysis of variance, however, we know that it would help us make better predictions if we knew to what part of the Archaic the projectile point pertained. If we knew that the point was Late Archaic, the best prediction would be the mean of the Late Archaic subsample. This prediction would more often be closer to the real weight than the prediction based on the overall sample mean. It is in this sense that we can say that knowing the period helps us to predict the projectile point weight. (We could, of course, reverse direction and predict period from weight. It is a little more complicated to phrase, and so we don't usually find it convenient to speak that way, but the relationship is symmetrical in that sense.)

If there were no relationship between projectile point weight and period, then knowing one would not help us predict the other at all. Looked at from this viewpoint, the significance question then becomes, "How likely is it that the relationship between projectile point weight and period that we observe in this sample is simply a consequence of sampling vagaries?" Yet another way to put it would be, "How

Be Careful How You Say It

The following sentence provides a complete example of how the conclusions from the example analysis of variance might be stated: "The difference observed in mean weight of Early, Middle, and Late Archaic projectile points in the Cottonwood River valley has extremely high significance ($F = 17.93$, $p = 0.0000001$)." This tells the reader what you concluded in a meaningful way; it says what significance test was used (because the F ratio is the result of an analysis of variance); and it gives the resulting statistic together with the significance level or associated probability. It would *not* be adequate simply to say, "The difference observed in mean weights of Early, Middle, and Late Archaic projectile points in the Cottonwood River valley is significant." This latter statement is not exactly incorrect, but it is certainly incomplete. It fails to specify what significance test was used, and it gives no information whatever about *how* significant the results were. It perpetuates the not very useful idea that being significant (like being pregnant) is a clearcut "yes" or "no" condition.

If what we are interested in is more easily framed in terms of the relationship between two variables, then there is yet a different way to phrase the overall conclusion to be drawn from the example analysis of variance: "For Archaic projectile points from the Cottonwood River valley, the relationship between weight and period has extremely high significance ($F = 17.93$, $p = .0000001$)."

One subtlety of reporting significance probabilities from computer output is to recognize what it means if your statpack reports a probability of 0.000. This does not mean absolute certainty. It only means a probability less than 0.0005, since anything greater than or equal to 0.0005 would round off to 0.001 and anything less than 0.0005 would round off to 0.000. Your program may enable you to ask it to show results to more decimal places so that you can see what the probability really is. If not, it is better to say that the probability is less than 0.0005 instead of saying that the probability is 0.000.

likely is it that we would select a sample of this size with this strong a relationship between weight and period from a population of projectile points in which the two variables were unrelated?" The analysis of variance answers this question with the F ratio and its associated probability. In our example, the answer to either question is "Extremely unlikely," corresponding to only one chance in ten million.

When the question we wish to ask is most naturally framed as one of relationship between two variables rather than differences between populations, then the analysis of variance can provide a convenient single answer to the question. When the question is more naturally framed as one of differences between populations, then the approach by way of estimating means for the different populations and attaching error ranges to the estimates (Fig. 13.1) is likely to be much more direct and informative.

ASSUMPTIONS AND ROBUST METHODS

Estimating population means and attaching error ranges to them is strongly affected by outliers. This problem can be corrected by estimating the trimmed mean and attaching an error range to it, as discussed in Chapter 9, and these estimated means and error ranges can be represented in a bullet graph. It all works exactly the same way as for making estimates from single samples, no matter how many subsamples are being compared. Each subsample is simply treated as an independent sample from which to estimate a population trimmed mean. If the trimmed mean is estimated from one subsample, however, the trimmed mean must be estimated from all subsamples. Comparing a trimmed mean to a regular mean is a comparison of apples and oranges. Estimating population means or trimmed means is, of course, of dubious value if the subsamples have asymmetrical shapes. Estimating medians may make more sense, and error ranges for the estimates can be determined with the bootstrap. The estimated means and their error ranges could be presented graphically in bullet graphs or as notched box-and-dot plots.

Analysis of variance assumes that the samples are drawn from populations with normal shapes, and that the spreads of the populations are approximately equal. Means of checking the validity of these assumptions, relying largely on stem-and-leaf plots and box-and-dot plots, were discussed at the beginning of the example analysis above. These assumptions will be recognized as precise parallels to the assumptions of the two-sample t test. If the spreads in the subsamples are very different, then that is, in itself, an indication that they did not come from identical populations. If the shapes of the subsamples are very asymmetrical, then a transformation that produces reasonably symmetrical shapes for all subsamples can be applied before going ahead with analysis of variance.

If the subsamples to be compared contain outliers, the analysis of variance can be based on the trimmed means and trimmed standard deviations, as discussed in Chapters 2 and 3. Few computer programs provide this as an option in analysis of variance, but it is not difficult to use most statistics packages to help you arrive at the trimmed mean and trimmed standard deviation for each subsample. Once these figures have been obtained, you have information analogous to that in Table 13.2, which you can use to calculate the final steps in the analysis of variance by hand as discussed in the text, simply using the trimmed mean and the trimmed standard deviation squared wherever the regular mean and the regular standard deviation squared are called for. (You will, of course, then need to go find an F table to look up the probability associated with the statistic you produce. Many statistics books contain this table.)

Table 13.4. Data on House Floor Area for Five Sites Occupied During the Early, Middle, and Late Neolithic Near Heiligenstadt

Floor area (m^2)	Site	Neolithic Subperiod	Floor area (m^2)	Site	Neolithic Subperiod	Floor area (m^2)	Site	Neolithic Subperiod
19.00	Hlg001	Early	16.83	Hlg004	Early	18.66	Hlg002	Early
16.50	Hlg004	Middle	16.43	Hlg004	Middle	18.36	Hlg004	Middle
16.10	Hlg002	Late	13.04	Hlg002	Late	16.07	Hlg005	Late
19.20	Hlg001	Late	21.14	Hlg003	Middle	17.17	Hlg002	Middle
15.20	Hlg005	Middle	18.24	Hlg005	Early	17.17	Hlg003	Late
20.40	Hlg001	Middle	17.34	Hlg002	Late	20.47	Hlg003	Late
16.40	Hlg002	Early	14.84	Hlg004	Middle	23.57	Hlg003	Late
16.40	Hlg002	Late	17.34	Hlg005	Middle	22.77	Hlg001	Middle
16.40	Hlg002	Middle	21.64	Hlg001	Early	22.77	Hlg003	Late
15.40	Hlg005	Early	15.74	Hlg005	Early	15.87	Hlg005	Late
20.60	Hlg001	Middle	19.84	Hlg001	Middle	15.08	Hlg005	Middle
17.20	Hlg004	Middle	22.99	Hlg003	Early	18.28	Hlg001	Early
19.90	Hlg003	Late	15.94	Hlg002	Middle	15.78	Hlg004	Late
22.01	Hlg001	Late	23.05	Hlg003	Middle	16.98	Hlg004	Late
21.11	Hlg003	Early	24.15	Hlg001	Early	20.58	Hlg001	Early
16.51	Hlg002	Early	20.35	Hlg003	Middle	16.08	Hlg004	Early
22.71	Hlg003	Middle	18.95	Hlg004	Early	21.68	Hlg003	Middle
20.81	Hlg001	Late	16.85	Hlg002	Middle	15.09	Hlg005	Late
15.81	Hlg005	Late	19.95	Hlg003	Early	17.79	Hlg004	Late
16.52	Hlg004	Early	20.16	Hlg001	Early	17.09	Hlg004	Late
21.12	Hlg003	Late	19.16	Hlg003	Middle	21.69	Hlg001	Middle
18.22	Hlg001	Late	17.66	Hlg004	Early	21.69	Hlg001	Late
23.22	Hlg003	Early	15.26	Hlg001	Middle	20.69	Hlg003	Early
16.32	Hlg005	Late	16.26	Hlg005	Middle	24.99	Hlg003	Early
16.13	Hlg005	Early	19.46	Hlg002	Late			
15.33	Hlg002	Early	15.46	Hlg005	Early			

PRACTICE

You are interested in investigating variability in group mobility, which you think is related to the size of the house that a family builds. You have excavated a series of Neolithic houses at five different sites near Heiligenstadt. Each site is in a different environmental setting, but each was occupied through all three parts of the Neolithic that you can identify: Early, Middle, and Late. The information is given in Table 13.4. A long Oktoberfest recess in your field seasons provides ample opportunity for deep consideration of issues of sampling bias, and you decide that you will use the sample of house floors from each site as a random sample from a much larger and vaguely defined population consisting of all house floors from environmental settings like that of the site in question. Likewise, you will use the sample of house floors from Early, Middle, and Late Neolithic as a random sample from the large and vaguely defined population of all house floors of that period.

1. Estimate the mean house floor area in each of the five environmental settings represented by the five different sites. Draw a bullet graph comparing these five populations in regard to estimated mean house floor area with error ranges for 80%, 95%, and 99% confidence levels. Does it appear that house sizes were different in different environmental settings? Summarize what you can conclude from your graph in one or two clear sentences.

2. Perform an analysis of variance to evaluate the relationship between house floor area and site based on this sample of 76 house floors. Does it appear that there is a relationship between environmental setting and house size? State the results of your analysis in one clearly worded sentence.

3. Estimate the mean house floor area for the region in each part of the Neolithic period. Draw a bullet graph comparing the Early, Middle, and Late Neolithic in regard to house size with error ranges for 80%, 95%, and 99% confidence levels. Does it appear that house size changed through time? Summarize what you can conclude from your graph in one or two clear sentences.

4. Perform an analysis of variance to evaluate the relationship between house size and period based on this sample of 76 house floors. Does it appear that there is a relationship between house size and period? State the results of your analysis in one clearly worded sentence.

Chapter 14
Comparing Proportions of Different Samples

Sometimes we have a sample divided into subsamples as in the example in Chapter 13, but the comparison we wish to make between the subsamples concerns not the mean of some measurement but rather another set of categories. Such a comparison can be approached by estimating population proportions from the various subsamples and attaching error ranges to the estimates. Then the estimated population proportions with their error ranges can be compared to each other with a bullet graph just as we did for means in Chapter 13.

COMPARISON WITH ESTIMATED PROPORTIONS AND ERROR RANGES

Table 14.1 provides some information about the quantities of sherds of two different vessel forms (bowls and jars) found at two sites (San Pablo and San Pedro). After carefully considering issues of sampling bias we decide that the methods by which these surface collections were made allow us to treat them as if they were random samples from the large and vaguely defined populations consisting of all the sherds at each site. We calculate the proportions of bowl and jar sherds in each sample and use these proportions as estimates of the corresponding population proportions, attaching error ranges to them on the basis of their standard errors, as discussed in Chapter 11. The estimate for the San Pablo site is 60% bowl sherds and 40% jar sherds, with a standard error of 9% for both. The estimate for the San Pedro site is 45% bowl sherds and 55% jar sherds, with a standard error of 8% for both.

R.D. Drennan, *Statistics for Archaeologists*, Interdisciplinary Contributions
to Archaeology, DOI 10.1007/978-1-4419-0413-3_14,
© Springer Science+Business Media, LLC 2004, 2009

Figure 14.1. Comparison of bowl and sherd proportions at the San Pablo and San Pedro sites..

These results are illustrated with a bullet graph in Fig. 14.1. Only the proportions of bowl sherds are graphed, since the bullet graph for jar sherds would show exactly the same contrast between the two sites in reverse. Based on these samples, we would say that the San Pablo site has a higher proportion of bowl sherds than the San Pedro site. Our confidence in this statement, however, would not be very high. Comparing the error ranges for different levels of confidence reveals that the estimated proportion for the San Pedro site falls well within the 99% confidence error range for the San Pablo site, and vice versa. Thus our confidence that our samples actually reflect a difference between the two sites (as opposed to reflecting just the vagaries of sampling) is less than 99%. Continuing the comparison, we note that the estimated proportion for the San Pedro site also falls within the 95% confidence error range for the San Pablo site and vice versa. Thus our confidence that our samples actually reflect a difference between the two sites is even less than 95%. The proportion for the San Pedro site does, however, fall outside the 80% confidence error range for the San Pablo site and vice versa. Thus our confidence that the observed difference reflects something more than just sampling vagaries is somewhere between 80% and 95% – moderate but not very high confidence. The difference might well be strong enough to be meaningful (a difference between 45% and 60%), but the risk that the difference might reflect nothing more than the chance variation between two relatively small samples from identical populations is higher than we might like.

COMPARISON WITH CHI-SQUARE

We first approached the comparison of a measurement between two or more samples by estimating population means (Chapters 12 and 13) and then turned to significance tests that boiled the entire comparison down to a single probability value (the *t* test

Table 14.1. Sherds of Different Vessel Forms from the San Pablo and San Pedro Sites

	Bowl sherds	Jar sherds	Total
San Pablo	18	12	30
San Pedro	18	22	40
Total	36	34	70

Table 14.2. Row Proportions of Sherds of Different Vessel Forms from the San Pablo and San Pedro Sites

	Bowl sherds	Jar sherds	Total
San Pablo	60.0%	40.0%	100.0%
San Pedro	45.0%	55.0%	100.0%
Average	51.4%	48.6%	100.0%

and analysis of variance). We have now, in similar fashion, approached the comparison of a set of categories between two samples by estimating population proportions. In this instance, too, there is a significance test that sums the entire comparison up in a single probability value. It is the chi-square test, named after the statistic that it produces, χ^2, represented by the Greek letter chi. The chi-square test works for any number of categories into which the overall sample is divided and for any number of categories for which proportions are calculated. Thus, for proportions, unlike means, there is no division between the two-sample case, where we used the t test to compare measurements, and the multiple-sample case, where we used analysis of variance for the same purpose.

Table 14.1 is easily recognized as the kind of table we worked with in Chapter 6. It seems natural to look at this table in terms of row proportions, because the rows are the two sites and it is the two sites that we want to compare to each other to investigate whether, for example, a difference in activities between the two sites might be reflected in different proportions of ceramic vessel forms. This is what we have, in fact, already been doing in comparing bowl proportions between the two sites. Table 14.2 provides these row proportions. We can see that the San Pablo site has a higher-than-average proportion of bowls, while the San Pedro site has a lower-than-average proportion of bowls – just what we concluded from Fig. 14.1. We could represent these departures from average with bar graphs as we did in Chapter 6, but this is such a simple comparison that it hardly seems necessary.

Chi-square is based on an assessment of these departures from average. This is accomplished by constructing a table of *expected* values to compare with the table of *observed* values (Table 14.1). If the average proportion of bowl sherds is 51.4%, as indicated in Table 14.2, then we would, in some sense, expect both the San Pablo and San Pedro sites to have 51.4% bowls. For the San Pablo site, this means 51.4% of 30 sherds, or 15.42 bowl sherds. For the San Pedro site, 51.4% of 40 sherds is 20.56 bowl sherds. Correspondingly, we would expect both sites to have 48.6% jar sherds. These expected values are shown in Table 14.3.

**Table 14.3. Expected Number of Sherds of Different Vessel
Forms from the San Pablo and San Pedro Sites**

	Bowl sherds	Jar sherds	Total
San Pablo	15.42	14.58	30
San Pedro	20.56	19.44	40
Total	36	34	70

Notice that the row and column totals (known together as the *marginal totals*) stay the same (allowing for rounding error) in the table of expected values as they were in the table of observed values. Indeed, it is the constant marginal totals upon which the expected values are based. The short cut for computing the expected values, in fact, is to multiply the row total corresponding to a particular cell by the column total corresponding to that cell and divide by the grand total for the table. For example, to obtain the expected number of bowl sherds at the San Pablo site, we could multiply the row total for that cell (30) by the column total for that cell (36) and divide by the grand total (70) to obtain 15.43 – exactly what we obtained from the row proportions (allowing for rounding error). We arrive at the same expected values for the table no matter whether we use row proportions, column proportions, or multiplication of marginal totals. This table of expected values provides the basis for a summary statistic, χ^2.

The χ^2 statistic is really very like a standard deviation, in that it involves calculating deviations, squaring them, and summing them up. The deviations, however, instead of being deviations from the mean, as they are for the standard deviation, are observed deviations from expected values:

$$\chi^2 = \sum \frac{(O_i - E_i)^2}{E_i}$$

where O_i = the observed value for the ith cell of the table, and E_i = the expected value for the ith cell of the table.

Our example is what is often referred to as a *two-by-two table* because it has two rows and two columns. There are, therefore, four cells. We thus calculate the quantity $(O_i - E_i)^2 / E_i$ for each of the four cells and sum up the four values:

$$\chi^2 = \frac{(18 - 15.42)^2}{15.42} + \frac{(12 - 14.58)^2}{14.58} + \frac{(18 - 20.56)^2}{20.56} + \frac{(22 - 19.44)^2}{19.44}$$
$$= 0.4317 + 0.4565 + 0.3188 + 0.3371$$
$$= 1.5441$$

This value, $\chi^2 = 1.5441$, is then looked up in Table 14.4 to determine the associated probability. One need only determine the appropriate number of degrees of freedom, which for χ^2 is the product of one less than the number of rows in the table times one less than the number of columns in the table. Since the table in our example

Table 14.4. The Chi-Square Distribution

Confidence	50%	80%	90%	95%	98%	99%	99.9%
	.5	.8	.9	.95	.98	.99	.999
Significance	50%	20%	10%	5%	2%	1%	0.1%
	.5	.2	.1	.05	.02	.01	.001
Degrees of freedom							
1	.455	1.642	2.706	3.841	5.412	6.635	10.827
2	1.386	3.219	4.605	5.991	7.824	9.210	13.815
3	2.366	4.642	6.251	7.815	9.837	11.341	16.268
4	3.357	5.989	7.779	9.488	11.668	13.277	18.465
5	4.351	7.289	9.236	11.070	13.388	15.086	20.517
6	5.348	8.558	10.645	12.592	15.033	16.812	22.457
7	6.346	9.803	12.017	14.067	16.622	18.475	24.322
8	7.344	11.030	13.362	15.507	18.168	20.090	26.125
9	8.343	12.242	14.684	16.919	19.679	21.666	27.877
10	9.342	13.442	15.987	18.307	21.161	23.209	29.588
11	10.341	14.631	17.275	19.675	22.618	24.725	31.264
12	11.340	15.812	18.549	21.026	24.054	26.217	32.909
13	12.340	16.985	19.812	22.362	25.472	27.688	34.528
14	13.339	18.151	21.064	23.685	26.873	29.141	36.123
15	14.339	19.311	22.307	24.996	28.259	30.578	37.697
16	15.338	20.465	23.542	26.296	29.633	32.000	39.252
17	16.338	21.615	24.769	27.587	30.995	33.409	40.790
18	17.338	22.760	25.989	28.869	32.346	34.805	42.312
19	18.338	23.900	27.204	30.144	33.687	36.191	43.820
20	19.337	25.038	28.412	31.410	35.020	37.566	45.315
21	20.337	26.171	29.615	32.671	36.343	38.932	46.797
22	21.337	27.301	30.813	33.924	37.659	40.289	48.268
23	22.337	28.429	32.007	35.172	38.968	41.638	49.728
24	23.337	29.553	33.196	36.415	40.270	42.980	51.179
25	24.337	30.675	34.382	37.652	41.566	44.314	52.620
26	25.336	31.795	35.563	38.885	42.856	45.642	54.052
27	26.336	32.912	36.741	40.113	44.140	46.963	55.476
28	27.336	34.027	37.916	41.337	45.419	48.278	56.893
29	28.336	35.139	39.087	42.557	46.693	49.588	58.302
30	29.336	36.250	40.256	43.773	47.962	50.892	59.703

(Adapted from Table 14 in *Tables for Statisticians* by Herbert Arkin and Raymond R. Colton (New York: Barnes and Noble, 1963)

has two rows and two columns, the number of degrees of freedom is $1 \times 1 = 1$. Using the first row in the table, then, for one degree of freedom, we see that the χ^2 value of 1.544 falls between the table values of 0.455 and 1.642. Thus the associated probability is between 50% and 20%. As with other significance tests, this is the probability that the differences we observe (in this case between the two sites in regard to proportions of sherds of different vessel forms) are a consequence of

---

Degrees of Freedom

In the tables on which the chi-square statistic is based, the term *degrees of freedom* makes some intuitive sense. There are, of course, numerous ways to fill cell values into a table so that they add up to a given set of marginal totals. In a two-by-two table, however, once a single cell value has been filled in, the other three cell values are determined because there is only one value for each of the other three cells that will make the given marginal totals add up correctly. This is, in a sense, the one degree of freedom that a two-by-two table has. For, say, a three-by-four table, there are six degrees of freedom (one less than the number of rows times one less than the number of columns), and it takes six cell values to completely determine such a table for a given set of marginal totals. (Try this out on paper, and you'll soon see just how it works. There is no set of five cells or fewer in a three-by-four table whose values will completely determine what the rest of the cell values must be to produce a given set of marginal totals. It takes six.)

Thinking back to calculations of standard deviations in sample batches of numbers and to the use of the t table reveals a related principle. For the t table, the number of degrees of freedom is one less than the number in the sample. If a sample batch has a given mean, then it is necessary to establish what all the numbers but one are before the last number is constrained to a single value. There is, of course, much more mathematical logic to this notion, but degrees of freedom in using the t table, in using the chi-square table, and, for that matter, dividing by $(n-1)$ in calculating the standard deviation of a sample are related to this notion.

---

the vagaries of sampling – that is to say, the probability that we could select two samples with proportions as different as these from parent populations having identical proportions. The chi-square test, then, is designed to answer the question, "How likely is it that we could select samples with proportions of bowl and jar sherds as different as these if the two sites did not really differ in regard to bowl and jar sherd proportions?"

In this example, our answer to the question is that there is somewhere between a 50% and a 20% risk that we could select samples as different as these if the two sites did not really differ in regard to bowl and jar sherd proportions. This is a high enough risk that our samples do not "really" indicate any difference between the two sites that we would not regard this evidence as much support for the notion of a difference in activities between the two sites. This is a slightly different conclusion than we came to from the bullet graph in Fig. 14.1. By looking at the bullet graph, we decided we would have between 80% and 95% confidence that there was a difference in bowl and jar sherd proportions between the two sites. A confidence level between 80% and 95% ought to translate into a significance probability between 20% and 5%, but the chi-square test gave us a significance probability between 50% and 20%.

Be Careful How You Say It

In conclusion to the chi-square example in the text, we can say "The difference between the San Pablo site and the San Pedro site with respect to proportions of bowl sherds and jar sherds is not very significant ($\chi^2 = 1.544$, $.50 > p > .20$)." This statement makes clear just what differences were investigated; it informs the reader what significance test was used, since χ^2 is the result of the chi-square test; and it provides the reader with the resulting statistic and its associated probability.

It would *not* be adequate to conclude this significance test simply by saying, "The San Pablo site and the San Pedro site do not differ significantly in proportions of bowl and jar sherds." In the first place, this latter statement does not tell the reader what significance test was used or provide its specific results. In the second place, it treats significance as a simple "yes" or "no" condition, which is at the least an oversimplification. On this last score, the inadequate statement even tends to mislead. The χ^2 value obtained (1.544) actually falls fairly close to the 20% significance column. Interpolating from the table, then, the actual probability must be only slightly greater than 20%. Put another way, the confidence we have that the two sites actually differ in bowl and jar sherd proportions is somewhere near 80%. We should, then, be saying that there is almost an 80% chance that the differences observed between the two samples actually do reflect differences between the sites rather than just the vagaries of sampling. The risk is still substantial (slightly over 20%) that nothing more may be at work here than the random variation of samples, but it is certainly more likely that there actually are differences in bowl and jar sherd proportions between the two sites. The last thing we want to do on the basis of this significance test is to act as if we have established that the two sites have the same proportions of bowl and jar sherds. This is why it is worth communicating that the odds favor the conclusion that there *is* a difference between the sites, even though there remains a worrisomely large risk that this may not be the case.

Under the influence of the near-sacred 5% significance level for rejecting or failing to reject the null hypothesis (see Chapter 12), people are accustomed to characterizing significance levels around 5% as "high." When the significance level reaches 1% or less, it is common to characterize it as "very high." A significance level around 20%, as in the example in the text, would usually be called "low." While it may seem that we've gone all the way from "very high" to "low" while staying pretty much toward one end of the scale, it is reasonable to think in such terms because, once the significance level goes far above 20%, the risk that the samples differ only because of the vagaries of sampling is so great that the result merits little attention. A difference between samples that is "highly significant" corresponds to "high confidence" that there is a difference. Note, though, that "high" significance corresponds to low associated significance probabilities (say, 5% or less), and "low" significance corresponds to high associated significance probabilities (say, around 20% or greater).

This happened because the two approaches are not just mirror image applications of the same principles. The error ranges in the bullet graph and the chi-square test are taking slightly different approaches, and it is not surprising that they produce slightly different results – yes, slightly different, because the two results are not really as different as they seem. If we look carefully at the bullet graph, we can see that the confidence we should have really is closer to 80% than to 95%. And if we look carefully at the chi-square table, we see that our result really is much closer to the 20% significance column than the 50% significance column. Thus, the bullet graph suggests slightly greater than 80% confidence, while the chi-square test suggests a significance slightly greater than 20%, so the two results do not in fact disagree by very much.

MEASURES OF STRENGTH

Just as in the other situations we have discussed, the significance and the strength of a result are different things. In this example, it is the significance level that gives us serious pause. There is somewhere around a 20% probability that we could select random samples and get the results that we got even if the two sites in fact had identical proportions of bowl and jar sherds. This is certainly far from reassuring. On the other hand, it is more likely that the difference observed between the two samples actually reflects a difference between the two sites. If this is the case, then the difference observed (15%) is probably strong enough to have a meaningful interpretation. Unlike the t test and analysis of variance, several specific measures of strength of results come along with the chi-square test's measure of significance.

One of the most flexible and easiest to calculate is Cramer's V:

$$V = \sqrt{\frac{\chi^2}{n(S-1)}}$$

where $n =$ the number of elements in the sample (that is, the grand total for the table), and $S =$ the number of rows or the number of columns in the table, whichever is smaller.

Thus, in our example,

$$V = \sqrt{\frac{1.544}{70(1)}} = 0.15$$

It can be shown that V ranges from zero to one. It takes on a value of zero when there is no difference at all between the observed values and the expected values, and it takes on a value of one when the difference between observed and expected values is as large as it can be. This latter would occur, for example, if the San Pedro site had only bowl sherds and the San Pablo site had only jar sherds. Thus, the closer V is to one, the stronger is the difference in proportions between the categories. (For a two-by-two table, V is the same as the difference between the observed proportions – here a difference of 15% between 60% and 45% for bowl sherds or 55% and 40% for jar sherds.)

For any table that has no more than two rows (or, alternatively, no more than two columns), the value of $S-1$ will always be 1, and this term will have no effect on the outcome. In this situation, V is equivalent to another measure of strength, called ϕ (the Greek letter phi, which in statistics is usually pronounced "fee"). The calculation of ϕ is quite simple: divide the χ^2 score by the grand total of the table and take the square root of the result. As long as the table is only two rows or only two columns ϕ is limited to a range between zero and one and is exactly the same thing as V. For tables with more than two rows or more than two columns, ϕ is not very useful because its range becomes open ended. V can be thought of as a modification of ϕ, expanding its utility to tables of any size. It is convenient simply to use V for tables of any size, and to recall that when someone refers to ϕ for a table of two rows or two columns it is the same thing as V.

THE EFFECT OF SAMPLE SIZE

Having obtained the results we did in the example chi-square test, we might decide that the possibility of differences between the two sites is intriguing, and we might want to explore it further. The very modest significance of the results, of course, was in part attributable to the fact that our samples were relatively small. (It is more likely that small samples will differ widely from the populations they are selected from than that large samples will. Thus the likelihood of getting a large difference between two small samples purely because of the vagaries of sampling is greater.) We might thus decide to seek larger samples of sherds from the two sites. Table 14.5 provides some imaginary results of seeking larger samples. Now there are exactly four times as many sherds, in exactly the same proportions as before. The strength of differences in proportions, then, remains the same (15%). Table 14.2 provides row proportions that are equally valid for the new result. The new expected values (Table 14.6) are, likewise, four times the old expected values (Table 14.3).

Calculating the χ^2 score, though, on the basis of the larger sample gives a very different result:

$$\chi^2 = \frac{(72-61.71)^2}{61.71} + \frac{(48-58.29)^2}{58.29} + \frac{(72-82.29)^2}{82.29} + \frac{(88-77.71)^2}{77.71}$$
$$= 1.7158 + 1.8165 + 1.2867 + 1.3626$$
$$= 6.1816$$

Table 14.5. A Larger Sample of Sherds of Different Vessel Forms from the San Pablo and San Pedro Sites

	Bowl sherds	Jar sherds	Total
San Pablo	72	48	120
San Pedro	72	88	160
Total	144	136	280

Table 14.6. Expected Numbers of Sherds of Different Vessel Forms from the San Pablo and San Pedro Sites with a Larger Sample

	Bowl sherds	Jar sherds	Total
San Pablo	61.71	58.29	120
San Pedro	82.29	77.71	160
Total	144	136	280

A χ^2 score of 6.1816 for one degree of freedom is associated with a significance level between 0.02 and 0.01. These results are highly significant. We might say, based on this sample, that we are between 98% and 99% confident that there are differences of bowl and jar sherd proportions between the San Pablo and San Pedro sites. The very different character of this conclusion from the one we reached before is attributable simply to sample size. Other things being equal, results from larger samples are more significant than results from smaller samples. The strength of the difference in proportions is still the same (a 15% difference between the sites in the proportion of bowl or jar sherds). And V continues to be 0.15.

For very small samples, only very strong results turn out to be significant. For larger samples even weaker results can be more significant. And for very large samples, even very weak results can have extremely high significance. Strength of results is most closely connected to meaning. It seems likely that the 15% difference in bowl sherd proportion between these two sites reflects some difference in the use of ceramic vessels at the two sites. Would a difference of only 5% have a similar meaning? Of 1%? Of 0.1%? At some point we would surely say that the difference in proportions was so weak that it meant little in terms of differences in ceramic vessel use. And yet we could certainly acquire a sample large enough to find even a tiny difference of 0.1% highly significant. It would clearly, however, not be worth the effort of acquiring such a large sample because it would not (at least in this regard) tell us anything useful. *Large samples are not necessarily more informative than smaller samples, because they may simply increase the statistical significance of results that are too weak to be meaningful.*

Precisely the same contrast between smaller and larger samples can, of course, be seen in estimates of population proportions like those illustrated in Fig. 14.1. The effect of the larger sample on that bullet graph would be to shorten the error bars substantially so that the difference in bowl proportions (which would remain the same) would be considerably more significant. If this does not seem intuitively sensible to you, try it out for yourself. Make a revised bullet graph with error bars calculated from the larger sample, and you will see just exactly how increasing the number of elements in the sample narrows in the error ranges for any given level of confidence.

DIFFERENCES BETWEEN POPULATIONS VERSUS RELATIONSHIPS BETWEEN VARIABLES

Just as analysis of variance can be thought of either as a study of differences between populations or as an investigation of relationship between variables, so can a chi-square test. In this instance, both variables are categorical. We would call the ones in the example analysis something like "site," with two categories (San Pablo and San Pedro), and "vessel form," also with two categories (bowls and jars). We have found an *association* between these two variables – an association of some strength but little significance. If we had framed the analysis as one investigating the relation between two variables rather than the difference between two populations, then we might conclude, "Vessel form proportions do differ somewhat from one site to the other, but there is little significance to the relationship between site and vessel form ($\chi^2 = 1.544, .50 > p > .20, V = 0.15$)."

As in Chapters 12 and 13, the bullet graph is an alternative to a significance test. For many purposes, the bullet graph in Fig. 14.1 would be the clearest and most straightforward way to present the observed differences between the San Pedro and San Pablo samples and the confidence we have in those observations. It would be statistical overkill to present such a bullet graph and then go on to present the results of a chi-square test. (It would indeed be a waste of time to calculate both.) They tell the same story. Pick the version that most serves the need at hand and use it.

Chi-square tests usually get more and more difficult to interpret meaningfully as the number of categories (and thus the number of rows and columns in the table) increases. With so many cells, it is usually only a few that show big differences between observed and expected values, and various techniques have been suggested for homing in on which specific cells in a large table really have important differences. They all boil down to comparing observed and expected values in some way, and for this reason it is common to include tables of observed and expected values to accompany chi-square results. It is often preferable to use bullet graphs in such a case, since they portray the categories individually, with the separate confidence/significance implications of each included in the graph.

ASSUMPTIONS AND ROBUST METHODS

The chi-square test does not involve means and standard deviations, so we have none of the worries associated with these indexes of level and spread in batches of true measurements. Thus assumptions concerning shapes are not made, and the issue of outliers simply does not arise. The principal concern about the chi-square test is that the sample be large enough for it to be a reliable approximation of the real probabilities. Many different rules of thumb can be found concerning this in different statistics texts. Some statisticians would like us not to use the chi-square

---

Statpacks

When framing a chi-square analysis as a test of relationship between two variables and when the data on the two categorical variables are organized as they are in Table 6.1, then computer statpacks are likely to be of considerable assistance in performing chi-square tests. If the data are already in the form of a table like that in Table 6.4 or Table 14.1, then calculation of the χ^2 score is probably more easily accomplished by hand. Most of the work is in counting up the numbers for the table, and it is under the heading of *cross tabulations* that many statpacks deal with chi-square. Just by way of comparison with the example we calculated in this chapter, a chi-square test performed on the data from Table 6.1 with a statpack indicates that the differences in proportions of incised and unincised ceramics from site to site are of moderate strength and little significance ($\chi^2 = 2.493$, $p = 0.29$, $V = 0.133$). The example from Table 14.1 in this chapter yields the same results when performed with a statpack that we already calculated by hand, except that, as usual with a statpack, the associated probability is calculated more precisely, $p = 0.214$, confirming the rough interpolation we made from Table 14.4 that the significance probability was between 20 and 50% but much closer to 20%.

---

test if any of the expected values in the table are less than 10. Others are much less conservative and are willing to accept chi-square tests based on tables with expected values as low as 1. A middle course, one that we will adopt here, is to insist that no expected value be less than 1 *and* that no more than 20% of the expected values be less than 5.

If these conditions are not met, and the table is a large one (that is, a table with many rows and/or many columns), it is often feasible to combine categories for one or both variables, so that there are fewer rows and/or fewer columns in the table. With the same number of cases divided among fewer cells, the expected values, of course, will be higher. Since we have adopted a relatively unconservative requirement for expected values, combining categories will ordinarily suffice to bring these expected values up to acceptable levels.

If a two-by-two table has such low expected values that the results of the chi-square test are unreliable, there is an alternative in the form of Fisher's exact test. This test is a direct calculation of the significance probability and there are no requirements at all about how large the expected values need be. Indeed, expected values do not even enter into its calculation. Fisher's method for calculating the exact significance probability for a two-by-two table is

$$p = \frac{(A+B)!(C+D)!(A+C)!(B+D)!}{N!A!B!C!D!}$$

where A = the observed frequency in the upper left cell of the two-by-two table, B = the observed frequency in the upper right cell of the two-by-two table, C = the observed frequency in the lower left cell of the two-by-two table, and D = the observed frequency in the lower right cell of the two-by-two table.

The mathematical use of the symbol ! may not be familiar. $X!$ (read "X factorial") means to multiply X sequentially by each positive integer less than X. For example, $5! = 5 \times 4 \times 3 \times 2 \times 1 = 120$. Or $9! = 9 \times 8 \times 7 \times 6 \times 5 \times 4 \times 3 \times 2 \times 1 = 362{,}880$. Calculating the probability for the example in Table 14.1 yields

$$p = \frac{(18+12)!(18+22)!(18+18)!(12+22)!}{70!\,18!\,12!\,18!\,22!} = 0.237$$

This is a calculation that most people will be willing to leave to their computers, but it does provide the exact significance probability for the example from Table 14.1 – a probability for which the chi-square test gives only an approximation. Most important, Fisher's exact test can be applied regardless of how low the expected cell values are, and when the numbers are small, the calculations are less formidable for those doing them by hand.

POSTSCRIPT: COMPARING PROPORTIONS TO A THEORETICAL EXPECTATION

Sometimes one arrives at a question in data analysis quite similar to the question we have been dealing with up to now in this chapter, but with one important difference. Perhaps our sample can be divided up into a set of categories and we know what we expect the proportions in those categories to be – based not on the proportions in another set of categories into which our sample can be divided, but rather on some different criterion. For example, suppose that we have results of regional survey in three different environmental settings as given in Table 14.7. Since most of the territory surveyed was in the river bottoms, we might expect to find most of the sites in that setting, other things being equal. As Table 14.7 makes clear, however, the proportions of sites in the three settings are quite different from what we might expect. But is our sample large enough to give us much confidence in these differences from our expectations?

At first glance, one might be tempted to answer this question with a chi-square test beginning as in Table 14.8. Thinking about that table, however, should make us pause. Table 14.8 does not really involve two variables that are two separate sets of categories for dividing up the same sample in two different ways. Instead it only involves one set of categories (the three environmental settings). Two different things have been divided up according to this one set of categories – the 38 sites and the $13.6\,\mathrm{km}^2$ of surveyed area. It makes no sense at all to add up 38 sites and $13.6\,\mathrm{km}^2$ and say that we have a sample of 51.6 (51.6 what?). Yet that is what we would be doing if we simply used the numbers in Table 14.8 to calculate χ^2.

Table 14.7. Regional Survey in Three Environmental Settings

Environmental setting	No. of sites	% of total sites	Area surveyed	
			km^2	% of total
Remnant levees	19	50.0	3.9	28.7
River bottoms	12	31.6	8.3	61.0
Slopes	7	18.4	1.4	10.3
Totals	38	100.0	13.6	100.0

Table 14.8. An *Incorrect* Way to Tabulate Observed Values from Table 13.7 for a Chi-Square Test

	No. of sites	Area surveyed (km^2)
Remnant levees	19	3.9
River bottoms	12	8.3
Slopes	7	1.4

In this example, what we have is a sample of 38 sites. If we are willing to treat this as a random sample of sites in the region, then we can compare the proportions of sites in different settings to the theoretical expectation based on how the territory surveyed to find the 38 sites was divided between the three different settings. One way to do this would be to follow the approach discussed in Chapter 11 for estimating the proportions of sites in different settings and for attaching error ranges to these estimates. This would tell us, for instance, that we can have 99% confidence that our sample of 38 sites came from a population in which 31.6% ± 20.3% of the sites were located in the river bottoms. (You can calculate this yourself, following the procedure in Chapter 11.) If prehistoric inhabitants showed no preference for any of these environmental settings, however, the proportion we would have expected here was 61.0%, since 61.0% of the territory is in this zone. This proportion is considerably higher than the top of the 99% confidence error range (51.9%), and so we can say that it is extremely unlikely that our sample came from a population with 61.0% sites in the river bottoms. Although we do find sites there, it seems that the prehistoric inhabitants showed something of an aversion to settling in the river bottoms. (Or possibly that recent sedimentation has covered more sites in the river bottoms than elsewhere, resulting in a particular failure to discover such sites on survey. This is a question of interpretation that statistical analysis of these numbers will not help us with.)

If we want to know exactly how unlikely it is that our sample came from a population with 61.0% sites in the river bottoms, we could perform a one-sample t test, as discussed in Chapter 12. This example is different from the example one-sample t test in Chapter 12 only in that there are three categories involved rather than just two. We could follow this approach to each of the three categories, determining whether the proportion estimated on the basis of our sample was greater or less than we would expect and how likely it was that the difference observed could be attributed to the amount of random variation ordinarily seen in samples the size of ours. This would lead to a specific discussion of settlement preferences (or apparent lack thereof) in regard to each of the three environmental settings.

We might want to treat the issue in a more comprehensive way, however, focusing not on each individual category, but asking the more general question, "How likely is it that this entire sample of 38 sites came from a population of sites in which there was no preference for locating sites in any particular environmental setting?" We can use a chi-square test to answer this question, but not in the way indicated in Table 14.8. Instead, we use the information we have to determine expected numbers of sites in each environmental setting as in Table 14.9. Since 28.7% of the surveyed area was on remnant levees, we might expect 28.7% of the 38 sites found (10.9 sites) to be on remnant levees, and so on. We now have a one-variable tabulation – observed and expected values for three categories. We can use these observed and expected values to calculate χ^2 just as before:

$$\chi^2 = \sum \frac{(O_i - E_i)^2}{E_i}$$

$$\chi^2 = \frac{(19 - 10.9)^2}{10.9} + \frac{(12 - 23.2)^2}{23.2} + \frac{(7 - 3.9)^2}{3.9}$$

$$= 6.0192 + 5.4069 + 2.4641$$

$$= 13.8902$$

Since there is only one row in this table (or one column – it makes no difference whether the table is vertical or horizontal), the number of degrees of freedom is one less than the number of categories. Here, there are three categories, so two degrees of freedom. A value of 13.8902 for χ^2 is just beyond the rightmost column of Table 14.4 in the row for two degrees of freedom. The rightmost column is for significance of .001. We could thus conclude, "It is extremely unlikely that this sample of sites was selected from a population in which sites were evenly distributed across environmental settings ($\chi^2 = 13.8902$, $p < .001$)." Or, "The difference between our survey results and the expected results was very highly significant ($\chi^2 = 13.8902$, $p < .001$)."

**Table 14.9. Observed and Expected Numbers of Sites
for Chi-Square Test**

	Area surveyed	No. of sites	
		Exp.	Obs.
Remnant levees	28.7%	10.9	19
River bottoms	61.0%	23.2	12
Slopes	10.3%	3.9	7
Totals	100.0%	38	38

PRACTICE

1. You have made surface collections at the Granger and Rawlins sites. Both collections include the same kinds of pottery, and you want to investigate whether the two sites differ in regard to the proportions of different pottery types. At the Granger site, you collected 162 sherds of the type Serengeti Plain, 49 sherds of the type Mandarin Orange, and 57 sherds of the type Zane Gray; from the Rawlins site you have 40 sherds of Serengeti Plain, 43 sherds of Mandarin Orange, and 49 sherds of Zane Gray. After considering possible sampling biases, you decide to use the collections as random samples from the populations consisting of all the sherds in each site. Estimate the proportions of the three pottery types at each site. Draw a bullet graph comparing the estimated proportions for the two sites with error bars for the 80%, 95%, and 99% confidence levels. (Think carefully about how to arrange the graph so that the error ranges you want to compare to each other are most easily compared.) How confident are you that the two sites differ in regard to proportions of ceramic types? Summarize the conclusions of your graphical comparison in one or two sentences.
2. Approach the issues raised in Question 1 by evaluating the strength and significance of the association between the variables site and pottery type. Summarize your results in one sentence. How do these results compare with those obtained in Question 1? What are the advantages and disadvantages of approaching these issues with a chi-square test rather than by estimating population proportions?

Table 14.10. Temper and Surface Finish for Sherds from the Opelousas Site

Temper	Surface	Temper	Surface	Temper	Surface
Sand	Red	Shell	Plain	Shell	Red
Sand	Red	Sand	Plain	Shell	Red
Sand	Red	Shell	Red	Sand	Plain
Shell	Plain	Shell	Plain	Sand	Plain
Sand	Red	Sand	Red	Sand	Red
Sand	Plain	Shell	Plain	Sand	Red
Sand	Red	Shell	Red	Sand	Plain
Shell	Plain	Sand	Red	Sand	Red
Shell	Red	Shell	Red	Shell	Plain
Shell	Red	Shell	Red	Sand	Red
Sand	Plain	Sand	Plain	Sand	Red
sand	Red	Sand	Red	Sand	Red
Sand	Red	Shell	Plain	Shell	Red
Sand	Plain	Shell	Red	Shell	Plain
Sand	Plain	Sand	Red	Shell	Red
Shell	Plain	Sand	Plain	Shell	Plain
Shell	Plain	Sand	Plain		
Shell	Red	Shell	Plain		

3. From the Opelousas site you have recovered a pitifully small collection of eroded sherds. You can't tell much about them except that some are tempered with shell and some with sand and that some were finished with a red slip while others have plain surfaces. The complete data are given in Table 14.10. Investigate the statistical significance and strength of any association between temper material and surface finish with the sample that you have. Summarize the meaning of your results in one clearly worded sentence.

Chapter 15
Relating a Measurement Variable to Another Measurement Variable

In Chapters 12 and 13 we investigated the relationship between a measurement variable and a categorical variable. We took two approaches to this task. The first was to estimate population means for the measurement variable in each of the categories of the categorical variable and attach error ranges to those estimates. The second approach was to use either a two-sample t test (if only two categories were involved) or an analysis of variance (if more than two categories were involved). In Chapter 14 we investigated the relationship between two categorical variables. Once again we took two approaches. The first was to estimate population proportions for one of the variables in each of the categories of the other variable and attach error ranges to those estimates. The second approach was to use a chi-square test to evaluate significance and Cramer's V to evaluate strength of association. There remains only to investigate the relationship between two measurement variables to complete all the possible combinations, and that is the subject of this chapter. We will see that one approach here is so powerful that we will not really consider alternative approaches.

Table 15.1 provides an example set of data consisting of observations on 14 known sites of the Oasis phase in the Río Seco valley. At each of the sites a systematic program of surface collection was undertaken to produce a sample of exactly 100 artifacts. After careful consideration of sources of bias we decide that we are willing to work with this sample of sites as if it were a random sample. Similarly considering sources of bias for the artifact collections, we decide we are willing to treat each as if it were a random sample of artifacts on the surface. Since each collection consists of 100 artifacts, the number of hoes in each is the percentage of hoes in the collection, and simultaneously our best approximation of the percentage

R.D. Drennan, *Statistics for Archaeologists*, Interdisciplinary Contributions
to Archaeology, DOI 10.1007/978-1-4419-0413-3_15,
© Springer Science+Business Media, LLC 2004, 2009

Table 15.1. Observations of Site Area and Number of Hoes in Collections of 100 Artifacts Made at Oasis Phase Sites in the Río Seco Valley

Site area (ha)	Number of hoes Per 100 artifacts	Site area (ha)	Number of hoes Per 100 artifacts
19.0	15	12.7	22
16.4	14	12.0	12
15.8	18	11.3	22
15.2	15	10.9	31
14.2	20	9.6	39
14.0	19	16.2	23
13.0	16	7.2	36

of hoes in each population (that is, the population of artifacts on the surface at each of the sites). In effect, of course, what we are dealing with here is a percentage, and percentages like this make perfectly suitable measurement variables to study in this way. What we want to investigate here is whether there is any relationship between the area of the site, as indicated by the extent of artifacts visible on the surface, and the number of hoes collected in the 100-artifact sample.

LOOKING AT THE BROAD PICTURE

As usual, drawing a plot that presents a picture of important aspects of the patterns to be observed is a good way to begin. The relationship between two measurement variables is best illustrated by a *scatter plot* (Fig. 15.1). Each x in the scatter plot represents one of the sites, and its position is determined according to the area of the site (in the horizontal direction) and the number of hoes in the collection of 100 artifacts (in the vertical direction).

Simple observation of this scatter plot begins to reveal something of the relationship between these two variables. The points toward the left of the graph (that is, with low values for site area) tend to fall fairly high on the graph (that is, they have high values for number of hoes). The points farther toward the right of the graph (that is, with high values for site area) tend to fall fairly low on the graph (that is they have low values for hoes). This suggests a pattern of larger sites having relatively fewer hoes per 100 artifacts and smaller sites having relatively more hoes per 100 artifacts.

In looking for patterns in scatter plots, especially if they are not very clear, it may sometimes help to look at groups of points separately and think of the levels in these sub-batches. For example, look at the points representing small sites (between 5 and 10 ha) in Fig. 15.1. There are only two such small sites, and both points fall very high on the graph, indicating that both sites have very high numbers of hoes. The center of this small batch of two sites is clearly quite high, perhaps around 37 hoes. In fact, these two smallest sites have the largest numbers of hoes of all the sites.

Figure 15.1. Scatter plot of number of hoes per 100 artifacts collected by site area.

Next look at the points in Fig. 15.1 that represent the middle-sized sites (between 10 and 15 ha). All these points fall lower on the graph than the points representing the small sites. The center of this batch of middle-sized sites falls lower than the center for the small sites, probably somewhere near 20 hoes. Clearly the middle-sized sites have fewer hoes than the small sites. Finally, look at the points representing the large sites (between 15 and 20 ha). The center of this batch is lower still, perhaps around 15 hoes. The same pattern emerges from this more detailed examination of the scatter plot that we saw on simple inspection: in general, the bigger the site, the smaller the number of hoes per 100 artifacts.

This detailed way of looking at the scatter plot suggests one way we could approach this problem. We could treat site area as three categories (small, medium, and large) and estimate the mean number of hoes per 100 artifacts in each of these categories. Then we could attach error ranges to these estimates and draw a bullet graph to illustrate the overall patterns. Or we could perform an analysis of variance – the other technique applicable to investigating the relationship between a measurement and a set of categories. Measurement variables can always be converted into a set of categories in this way, and sometimes it is useful to do so. There is, however, a much more powerful way to approach the investigation of the relationship between two measurements.

LINEAR RELATIONSHIPS

The easiest kind of relationship to describe between two measurements is a *linear*, or straight-line, relationship. Such a relationship is called linear because it is represented by a straight line on a scatter plot. Perhaps the simplest possible relationship between two measurements is when the one equals the other. If we let X represent

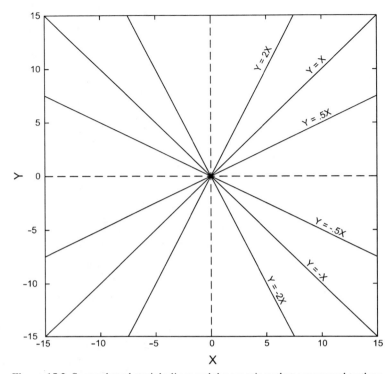

Figure 15.2. Some plotted straight lines and the equations that correspond to them.

one of the measurements and Y represent the other, then the relationship of equivalence is simply expressed by the equation $Y = X$. For any given value of X, there is a corresponding value of Y, which is determined easily by the equation. For example, when $X = 5$, then $Y = 5$; when $X = -10$, then $Y = -10$. The values of X and Y are plotted on the graph in Fig. 15.2. (By convention we always use the horizontal axis for X and the vertical axis for Y.) All the points representing pairs of X and Y values that satisfy the equation $Y = X$, lie along the line labeled $Y = X$ in Fig. 15.2 – a perfect straight-line relationship between X and Y.

The other lines in Fig. 15.2 also represent perfect straight-line relationships between X and Y. They are labeled with the corresponding equations. The positions of these lines can be determined experimentally. For example, the line that represents $Y = -2X$ is defined by all the points that satisfy the equation $Y = -2X$. These include $X = 5$, $Y = -10$; $X = -7$, $Y = 14$ and so on. The equations in Fig. 15.2 are algebraic expressions of relationships between two measurements and the lines on the graph are geometric expressions of the same relationships. Each equation comprises a complete description of the corresponding line. If this is at all unclear, it is a good idea to experiment on your own with some equations and their corresponding

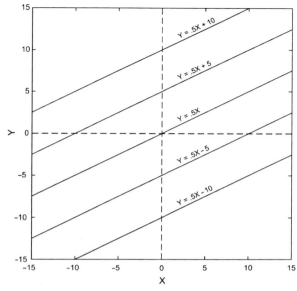

Figure 15.3. More plotted straight lines and the equations that correspond to them.

graphs. Make up some values for X; calculate the corresponding values for Y; and plot the points.

Comparison of the equations in Fig. 15.2 reveals one property of the relationship between equations and lines. If Y simply equals X multiplied by some number, then the relationship is represented geometrically by a straight line, and that straight line passes through the origin of the graph. (The origin is the point where $X = 0$ and $Y = 0$.) The number by which X is multiplied in the equation is called the *coefficient* of X, and it is this coefficient that governs the *slope* of the line. If the coefficient of X is positive, then the line rises as it moves from left to right. If the coefficient is negative, then the line falls as it moves from left to right. The larger the absolute value of the coefficient, the steeper the slope. That is, $Y = 2X$ has a steeper slope than $Y = .5X$; and $Y = -2X$ has a steeper slope than $Y = -.5X$. (In the equation $Y = X$, of course, the coefficient of X is 1.)

Figure 15.3 illustrates the other principal characteristic of straight lines on a graph – their positions relative to the origin. All the lines on the graph in Fig. 15.3 have the same slope – the *coefficient* of X is .5 in every case. They differ, however, in the degree to which they are offset from the origin. These equations differ only in having an additional term added to the product of X and its coefficient. The line corresponding to $Y = .5X + 5$ crosses the Y axis at the point where $Y = 5$. (This, of course, has to be true because X is 0 at the Y axis, and when $X = 0$ then $Y = 5$.) This additional term is called the Y *intercept* since it is the value of Y when X is 0, which is to say, the value of Y where the straight line crosses the Y axis.

Thus for any straight-line relationship between X and Y we can write an equation in the form

$$Y = bX + a$$

where b = the slope of the line, and a = the Y intercept, or value of Y where the line crosses the Y axis.

This specifies exactly what the relationship between X and Y is. It enables us to say, for a given value of X, what Y will be. By convention, we always take X as a given, and let Y's value depend on X. Thus X is the *independent variable* and Y is the *dependent variable*.

THE BEST-FIT STRAIGHT LINE

We have strayed rather far from the example where we wanted to investigate the relationship between site area and number of hoes collected per 100 artifacts. The point of the discussion of straight-line relationships, however, was to make clear exactly what kind of mathematical relationship we might expect to find between these two measurements. If the relationship between site area and number of hoes collected can be described reasonably accurately as a straight-line relationship, then we can characterize it in these terms. If, for example, the scatter plot in Fig. 15.1 had looked like Fig. 15.4 instead, we would find it quite easy to apply the principles of straight-line equations just discussed. The points in Fig. 15.4 do fall almost perfectly along a straight line, and an approximation of that line has been drawn on the graph. We could measure the slope of the line and determine the Y value of the point at

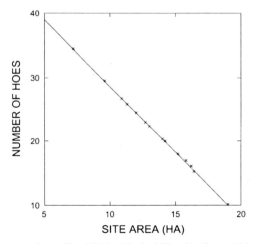

Figure 15.4. If the scatter plot in Fig. 15.1 had looked like this, it would have been easy to fit a straight line to the points.

which it crosses the Y axis and write an equation that specified the relationship between the two measurements algebraically.

The problem, of course, is that the points in the real scatter plot from our example data did not fall almost perfectly along a straight line. While the general pattern of declining numbers of hoes per 100 artifacts with increasing site area was clear, no straight line could be drawn through all the points. There are so many advantages to working with straight-line relationships, though, that it is worth trying to draw a straight line on Fig. 15.1 that represents the general trend of the points as accurately as possible – a *best-fit straight line*. The statistical technique for accomplishing this is *linear regression*.

The conceptual starting point for linear regression is to think exactly what criterion would determine which line, of all the possible straight lines we could draw on the scatter plot, would fit the points best. Clearly, we would like as many of the points as possible to lie as close to the line as possible. Since we take the values of X as given, we think of closeness to the line in terms of Y values only. That is, for a given X value, we think of how badly the point "misses" the line in the Y direction on the graph. These distances are called *residuals*, for reasons that will become clear later on.

We can explore the issue of residuals with the completely fictitious scatter plot of Fig. 15.4. Since the points in this scatter plot do fall very closely along a straight line, it is a bit easier to see good and bad fits. Fig. 15.5 illustrates a straight line that does not fit the pattern of points nearly so well as the one in Fig. 15.4. We can see that simply by inspection. We could put a finer point on just how bad the fit is by measuring the residuals, which are indicated with dotted lines in Fig. 15.5. The measurements, of course, would be taken vertically on the graph (that is, in the Y

Figure 15.5. A straight line that does *not* fit the points from Fig. 15.4 very well.

direction) and would be in terms of Y units (that is, in this example, in terms of numbers of hoes). The same operation could be performed algebraically as well. Since the line corresponds to a specific linear equation relating X and Y, we could use that equation to calculate, for each X value, the value of Y that would be "correct" according to the relationship the line represents. The difference between that "correct" value of Y and the actual value of Y would correspond to the graphical measurement of the residual. The residual and its measurement are shown for the leftmost point on the scatter plot in Fig. 15.5. This point falls 4.88 Y units above the straight line. The residual corresponding to this point, then, is 4.88. This means that this site actually had 4.88 more hoes per 100 artifacts collected than the value we would calculate *based on the straight line drawn in Fig. 15.5.*

We can easily see that the straight line in Fig. 15.5 could be adjusted so that it followed the trend of the points better by twisting it around a bit in a clockwise direction. If we did this, the dotted lines representing the residuals could all be shortened substantially. Indeed, we would put the line back the way it was in Fig. 15.4, and the residuals would all be zero or nearly zero. Thus we can see that minimizing residuals provides a mathematical criterion that corresponds well to what makes good sense to us from looking at the scatter plot. The better the fit between the straight line and the points in the scatter plot, the smaller the residuals are collectively.

The residuals amount to deviations between two alternate values of Y for a given value of X. There is the Y value represented by the straight line and there is the Y value represented by the data point. As usual in statistics, it turns out to be most useful to work not directly with these deviations but with the squares of the deviations. Thus, the most useful mathematical criterion is that *the best-fit straight line is the one for which the sum of the squares of all the residuals is least.* From this definition comes a longer name for the kind of analysis we are in the midst of: *least-squares regression.*

The core of the mathematical complexity of regression analysis, as might be expected, concerns how we determine exactly which of all the possible straight lines we might draw provides the best fit. Fortunately it is not necessary to approach this question through trial and error. Let's return to the general form of the equation for a straight line relating X and Y:

$$Y = bX + a$$

It can be shown mathematically that the following two equations produce values of a and b that, when inserted in the general equation, describe the best-fit straight line:

$$b = \frac{n \sum X_i Y_i - (\sum Y_i)(\sum X_i)}{n \sum X_i^2 - (\sum X_i)^2}$$

and

$$a = \overline{Y} - b\overline{X}$$

where n = the number of elements in the sample, X_i = the X value for the ith element, and Y_i = the Y value for the ith element.

Since the summations involved in the equation for b are complex, it is perhaps worth explaining the operation in detail. For the first term in the numerator of the fraction, $n\sum X_i Y_i$, we multiply the X value for each element in the sample by the Y value for the same element, then sum up these n products, and multiply the total by n. For the second term in the numerator, $(\sum Y_i)(\sum X_i)$, we sum up all n X values, sum up all n Y values, and multiply the two totals together. For the first term in the denominator, $n\sum X_i^2$, we square each X value, sum up all these squares, and multiply the total by n. And for the second term in the denominator, $(\sum X_i)^2$, we sum up all the X values and then square the total. Having arrived at a value for b, deriving the value of a is quite easy by comparison. We simply subtract the product of b times the mean of X from the mean of Y.

There are computational shortcuts for performing these cumbersome calculations, but in fact there is little likelihood that any reader of this book will perform a regression analysis without a computer, so we will not take up space with these shortcut calculations. Neither will we laboriously work these equations through by hand to arrive at the actual numbers for our example. This example has been performed the way everyone now can fully expect to perform a regression analysis, by computer. The point of including the equations here, then, is not to provide a means of calculation but instead to provide insight into what is being calculated and thus into what the results may mean.

PREDICTION

Once we have the values of a and b, of course, we can specify the equation relating X and Y and, by plugging in any two numbers as given X values, determine the corresponding Y values and use these two points to draw the best-fit straight line on the graph. If we do this for the data from Table 15.1, we get the result shown in Fig. 15.6. The values obtained in this regression analysis are

$$a = 47.802$$

and

$$b = -1.959$$

Thus the equation relating X to Y is

$$Y = -1.959X + 47.802$$

or

$$Number\ of\ hoes = (-1.959 \times Site\ area) + 47.802$$

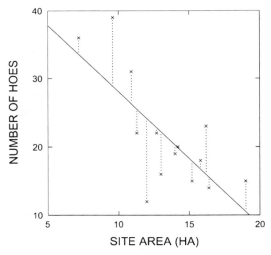

Figure 15.6. The best-fit straight line for the points from Fig. 15.1.

This equation literally enables us to "predict" how many hoes there will be per 100 artifacts collected if we know the site area. For example, if the site area is 15.2 ha, we predict

$$Y = (-1.959)(15.2) + 47.802 = 18.03$$

Thus, if the relationship between X and Y described by the regression equation holds true, a site with an area of 15.2 ha should yield 18.03 hoes in a collection of 100 artifacts. There actually was a site in the original data set with an area of 15.2 ha, and the collection of 100 artifacts from that site had 15 hoes. For this site, then, reality fell short of the predicted number of hoes by 3.03. Thus the residual for that site is −3.03, representing a bit of variation unpredicted or "unexplained" by the regression equation. (The name "residual" is used because residuals represent unexplained or leftover variation.) The prediction based on the regression equation is, however, a better prediction than we would otherwise be able to make. Without the regression analysis our best way to predict how many hoes would be collected at each site would be to use the mean number of hoes for all sites, or 21.57 hoes. This would have meant an error of 6.57 hoes in the case of the 15.2 ha site. In this instance, then, the regression equation has enabled us to predict how many hoes would be found on the basis of site area more accurately than we could if we were unaware of this relationship. This will not necessarily be true for every single case in a regression analysis, but it will be true on average.

Regression analysis, then, has helped us to predict or "explain" some of the variation in number of hoes collected per 100 artifacts. It has, however, still left some of the variation "unexplained." We do not know why the 15.2 ha site had 3.03 fewer hoes than we expected. The residuals represent this unexplained variation, a subject to which we shall return below.

HOW GOOD IS THE BEST FIT?

We know that the equation

$$Y = -1.959X + 47.802$$

represents the best-fit straight line for our example data, so that the sum of the squares of the residuals is the lowest possible (for straight-line equations). These residuals are shown with dotted lines in Fig. 15.6. We notice immediately that some of them are quite large. Although the best-fit straight line does help us predict or explain some of the variability in number of hoes collected, it clearly does not fit the data as well as we might have hoped. It would be useful for us to be able to say just how good a fit it is, and the very process of determining the best-fit straight line provides us with a way to do so. Since the best-fit straight line is the one for which the sum of the squares of the residuals is least, then the lower the sum of the squares of the residuals, the better the fit. The sum of the squares of the residuals becomes a measure of how well the best-fit straight line fits the points in the scatter plot.

The sum of the squares of the residuals, of course, can never be less than 0, because there will never be a negative number among the squared residuals that are summed. (Even negative residuals have positive squares.) The sum of the squares of the residuals will only be 0 when all the residuals are 0. This only happens when all the points lie exactly on the straight line and the fit is thus perfect. There is no fixed upper limit on the sum of the squares of the residuals, however, because it depends on the actual values taken by Y. It would be useful if we could determine this upper limit because then we would know just where, between the minimum and maximum possible values, a particular sum of squared residuals lay. We could then determine whether the best-fit straight line really was closer to the best of all possible fits (a value of zero for the sum of the squares of the residuals) or the worst of all possible fits (whatever that maximum value for the sum of the squares of the residuals might be). It turns out that the maximum value the sum of the squares of the residuals can have is the sum of the squares of the deviations of Y from its mean. (The sum of the squares of the deviations of Y from its mean is, of course, the numerator in the calculation of the variance of Y – that is, $\sum (y_i - \overline{Y})^2$.) Thus the ratio

$$\frac{(sum\ of\ the\ squares\ of\ residuals)}{\sum (y_i - \overline{Y})^2}$$

ranges from zero to one. Its minimum value of zero indicates a perfect fit for the best-fit straight line because it only occurs when all the residuals are zero. Its maximum value of one indicates the worst possible fit because it occurs when the sum of the squares of the residuals is as large as it can be for a given set of values of Y (that is, equal to $\sum (y_i - \overline{Y})^2$).

This ratio, then, enables us to say, on a scale of zero to one, how good a fit the best-fit straight line is. Zero means a perfect fit, and one means the worst possible fit. It is easier intuitively to use a scale on which one is best and zero is worst, so we customarily reverse the scale provided by this ratio by subtracting the ratio from one. (If this does not make intuitive good sense to you, try it with some numbers. For example, 0.2 on a scale from zero to one becomes 0.8 on a scale from one to zero.) This ratio, when subtracted from one, is called r^2, and

$$r^2 = 1 - \frac{(sum\ of\ the\ squares\ of\ residuals)}{\Sigma\left(y_i - \overline{Y}\right)^2}$$

The ratio, r^2, amounts to a ratio of variances. The denominator is the original variance in Y (omitting only the step of dividing by $n - 1$) and the numerator is the variance that Y has from the best-fit straight line (again omitting only the step of dividing by $n - 1$). Including the step of dividing by $n - 1$ would have no effect on the result since it would occur symmetrically in both numerator and denominator.

If the variation from the best-fit straight line is much less than the original variation of Y from its mean, then the value of r^2 is large (approaching one) and the best-fit straight line is a good fit indeed. If the variation from the best-fit straight line is almost as large as the original variation of Y from its mean, then the value of r^2 is small (approaching zero) and the best-fit straight line is not a very good fit at all. Following from this logic, it is common to regard r^2 as a measure of the *proportion of the total variation in Y explained by the regression*. This also follows from our consideration of the residuals as variation unexplained or unpredicted by the regression equation. All this, of course, amounts to a rather narrow mathematical definition of "explaining variation," but it is useful nonetheless within the constraints of linear regression. For our example, r^2 turns out to be 0.535, meaning that 53.5% of the variation in number of hoes per collection of 100 artifacts is explained or accounted for by site area. This is quite a respectable amount of variation to account for in this way.

More commonly used than r^2, is its square root, r, which is also known as *Pearson's r* or the *product-moment correlation coefficient* or just the *correlation coefficient*. We speak, then, of the *correlation* between two measurement variables as a measure of how good a fit the best-fit straight line is. Since r^2 ranges from zero to one, then its square root must also range from zero to one. While r^2 must always be positive (squares of anything always are), r can be either positive or negative. We give r the same sign as b, the slope of the best-fit straight line. As a consequence, a positive value of r corresponds to a best-fit straight line with a positive slope and thus to a positive relationship between X and Y, that is, a relationship in which as X increases Y also increases. A negative value of r corresponds to a best-fit straight line with a negative slope and thus to a negative relationship between X and Y, that is, a relationship in which as X increases Y decreases. The correlation coefficient r, then, indicates the direction of the relationship between X and Y by its sign, and it indicates the strength of the relationship between X and Y by its absolute value on a scale from zero for no relationship to one for a perfect relationship (the strongest possible). In our example, $r = -0.731$, which represents a relatively strong (although negative) correlation.

SIGNIFICANCE AND CONFIDENCE

Curiously enough, the question of significance has not arisen up to now in Chapter 15. The logic of our approach to relating two measurement variables has been very different from our approach to relating two categorical variables or one measurement variable and one categorical variable. Through linear regression, however, we have arrived at a measure of strength of the relationship, r, the correlation coefficient. This measure of strength is analogous to V, the measure of strength of association between two categorical variables. It is analogous to the actual differences between means of subgroups in analysis of variance as an indication of the strength of relation between the dependent and independent variables. We still lack, however, a measure of the significance of the relationship between two measurements. What we seek is a statistic analogous to χ^2 for two categorical variables, or t or F for a categorical variable and a measurement – a statistic whose value can be translated into a statement of how likely it is that the relation we observe is no more than the effect of the vagaries of sampling.

Much of our discussion about arriving at the best-fit straight line and providing an index of how good a fit it is centered on variances and ratios of variances. This sounds a great deal like analysis of variance, and indeed it is by calculating F as a ratio of variances that we arrive at the significance level in a regression analysis. In analysis of variance we had

$$F = \frac{s_B^2}{s_W^2} = \frac{SS_B / d.f.}{SS_W / d.f.}$$

which is to say

$$F = \frac{(sum\ of\ squares\ between\ groups / d.f.)}{(sum\ of\ squares\ within\ groups / d.f.)}$$

In regression analysis we have

$$F = \frac{(sum\ of\ squares\ explained\ by\ regression / d.f.)}{(sum\ of\ squares\ unexplained\ by\ regression / d.f.)}$$

This is equivalent to

$$F = \frac{r^2 / 1}{(1 - r^2) / (n - 2)}$$

In our example, $F = 13.811$, with an associated probability of 0.003. As usual, very low values of p in significance tests indicate very significant results. There are several ways to think about the probability values in this significance test. Perhaps the clearest is that this result indicates a probability of 0.003 of selecting a random sample with a correlation this strong from a population in which these two variables were unrelated. That is, there are only three chances in 1,000 that we could select a sample of 14 sites showing this strong a relation between area and number of hoes from a population of sites in which there was no relation between area and number of hoes. Put yet another way, there is only a 0.3% chance that the relationship we observe in our sample between site area and number of hoes reflects nothing more than the vagaries of sampling. If we are willing to treat these 14 sites as a random sample of Oasis phase sites from the Río Seco valley, then, we are 99.7% confident in asserting that larger Oasis phase sites in the Río Seco valley tend to have fewer hoes per 100 artifacts on their surfaces.

As usual, significance probabilities can be used to tell us how likely it is that the observation of interest in our sample (in this case the relationship between site area and number of hoes) does not actually exist in the population from which the sample was selected. We can also discuss regression relationships in terms of confidence, in a manner parallel to our earlier use of error ranges for different confidence levels. In this case, instead of an individual estimate ± an error range, it is useful to think of just what the relationship between the two variables is likely to be in the population from which our sample came. We know from the significance probability obtained in our example that it is extremely unlikely that there is no relationship at all between site area and number of hoes in the population of sites from which our 14 sites are a sample. The specific relationship expressed by the regression equation derived from analysis of our sample is our best approximation of what the relationship between site area and number of hoes is in the population. But, as in all our previous experience with samples, the specific relationship observed in the sample may well not be exactly the same as the specific relationship that exists in the population as a whole. Most likely the regression equation we would obtain from observing the entire population (if we could) would be similar to the one we have derived from analysis of the sample. It is less likely (but still possible) that the relationship in the population as a whole is rather different from the relationship observed in the sample. And, as

---

Be Careful How You Say It

We might report the results of the example regression analysis in the text by saying, "For Oasis phase sites in the Río Seco valley there is a moderately strong correlation between site area (X) and number of hoes per collection of 100 artifacts (Y) ($r = -.731$, $p = .003$, $Y = -1.959X + 47.802$)." This makes clear what relationship was investigated; it lets the reader know what significance test was used; it provides the results in terms of both strength and significance; and it states exactly what the best-fit linear relationship is. Like "significance," the word "correlation" has a special meaning in statistics that differs from its colloquial use. It refers specifically to Pearson's r and other analogous indexes of the relationship between two measurements. Just as "significant" should not be used in statistical context to mean "important" or "meaningful," "correlated" should not be used in statistical context to refer simply to a general correspondence between two things.

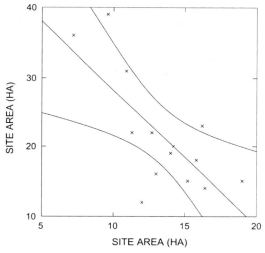

Figure 15.7. The best-fit straight line with its 95% confidence zone.

the significance probability has already told us, it is extremely unlikely (only three chances in 1,000) that there is no relationship at all between site area and hoes in the population.

This range of possible relationships that might exist in the population our sample came from, and their varying probabilities, can be depicted graphically as in Fig. 15.7. It is neither very practical nor very enlightening to discuss the calculation of the curves that delimit this 95% confidence region. In practice, it is almost unimaginable now to produce such a graph except by computer, so we will concentrate on what the graph tells us. The 95% confidence region, which includes the best-fit straight line for our sample in its center, depicts the zone within which we have 95% confidence that the best-fit straight line for the population lies. There is only a 5% chance that the best-fit straight line for the population of sites from which our sample of 14 was selected (if we could observe the entire population) would not lie entirely between the two curves. Determination of this confidence region, then, enables us to think usefully about the range of possible relationships between site area and number of hoes likely to exist in the population from which our sample came. Depiction of this confidence region relates to the significance probability in the same way that our previous use of error ranges for different confidence levels related to parallel significance tests.

ANALYSIS OF RESIDUALS

The regression analysis described in the example used above has enabled us to explain a portion of the variation in number of hoes per collection of 100 artifacts. One possible interpretation of these results is that larger settlements contained

Table 15.2. Hoes at Oasis Phase Sites in the Río Seco Valley: Predictions and Residuals

Site area (ha)	Number of hoes per 100 artifacts	Number of hoes predicted by regression on site area	Residual number of hoes
19.0	15	10.59	4.41
16.4	14	15.68	−1.68
15.8	18	16.86	1.14
15.2	15	18.03	−3.03
14.2	20	19.99	0.01
14.0	19	20.38	−1.38
13.0	16	22.34	−6.34
12.7	22	22.93	−0.93
12.0	12	24.30	−12.30
11.3	22	25.67	−3.67
10.9	31	26.45	4.55
9.6	39	29.00	10.00
16.2	23	16.07	6.93
7.2	36	33.70	2.30

larger numbers of craft workers and elite residents and fewer farmers. Thus hoes were scarcer in the artifact assemblages at the larger sites. (We would presumably have had something like this in mind in the first place or we would not likely have been interested in investigating the relationship between site area and number of hoes at all. We would also presumably have provided the additional evidence and argumentation necessary to make this a truly convincing interpretation.)

Since the regression analysis has explained part of the variation in number of hoes, it has also left another part of this variation unexplained. This unexplained variability is made specific in the form of the residuals. The 15.2-ha site that we discussed, for instance, actually had 3.03 fewer hoes than the regression analysis led us to expect, based on the size of the site. This 3.03 is its residual, or leftover variation. For each site there is, likewise, a residual representing how much the observed number of hoes differed from the predicted number of hoes. Table 15.2 provides the original data together with two new items. For each site, the number of hoes per collection of 100 artifacts predicted on the basis of the regression equation relating number of hoes to site area is listed. Then comes the residual for each site (that is the number of hoes actually collected minus the number predicted by the regression equation).

In examining the residuals, we note as expected that some sites had considerably fewer hoes than we predicted and some had substantially more than we predicted. We can treat these residuals as another variable whose relationships can be explored. In effect, the regression analysis has created a new measurement – the variation in number of hoes unexplained by site size. We can deal with this new measurement just as we would deal with any measurement we might make. We would begin to explore it by looking at a stem-and-leaf plot and perhaps a box-and-dot plot. We would be interested, for example, in the possibility of multiple peaks in this new batch of numbers. A two-peaked shape would suggest two distinct sets of sites, probably one with substantially more hoes than we would expect (given site size), and one with substantially fewer hoes than we would expect. We might be able to determine some other characteristics of these two groups of sites that helped us to understand why they deviated in such different ways from the number of hoes we would expect, given their size. If the shape is single peaked we might go on to explore the relationship between this new batch of measurements and other variables. For example, we might imagine that, in addition to site area reflecting the presence of nonfarming specialists, residents of sites in very fertile soils might dedicate themselves more intensively to farming than residents of sites in very poor soils. We might, then, investigate the relationship between our new measurement (the residuals from the regression analysis) and fertility of soils for each site.

Table 15.3 provides just such information about the productivity of soils – the estimated yield of maize (in kilograms per hectare) – at each of the 14 sites in the Río Seco valley. Examination of a stem-and-leaf plot reveals that both batches of numbers (the residuals and the soil productivity figures) are single peaked and symmetrical, so we can proceed to investigate whether sites that have more hoes than we would expect on the basis of their size are those located in more productive soils. Both variables are true measurements, so again the technique of choice is regression analysis. The scatter plot for these two variables (Fig. 15.8) suggests a strong positive relationship. Just as we expected, the sites on the more productive soils tend to have more hoes than expected, based on their size (positive residuals), and those on the less productive soils tend to have fewer hoes than expected, based on their size (negative residuals). The best-fit straight line looks to be quite a good fit, and the 95% confidence zone around it is tight. The regression analysis fully

Table 15.3. Residual Numbers of Hoes and Soil Productivity for Sites in the Río Seco Valley

Residual number of hoes	Soil productivity (kg of maize per ha)
4.41	1,200
−1.68	950
1.14	1,200
−3.03	600
0.01	1,300
−1.38	900
−6.34	450
−0.93	1,000
−12.30	350
−3.67	750
4.55	1,500
10.00	2,300
6.93	1,650
2.30	1,700

Figure 15.8. Scatter plot of residual number of hoes by soil productivity with best-fit straight line and 95% confidence zone.

confirms all these observations. The correlation is very strong and highly significant ($r = 0.923$, $p < 0.0005$). Since $r^2 = 0.852$, 85.2% of the variation in the residuals is explained by soil productivity.

The results of these two regression analysis are complementary and contribute cumulatively to our goal of explaining the variation in number of hoes at these sites. The first regression analysis (number of hoes by site area) showed that site area accounted for 53.5% of the variation in number of hoes, leaving 46.5% of the variation in number of hoes unexplained. It is that 46.5% of the variation left unexplained by the first regression that is encapsulated in the residuals. The second regression analysis (residual number of hoes by soil productivity) accounted for 85.2% of the variation in hoe residuals, which was in turn the 46.5% of the variation in number of hoes left unexplained by the first regression. This amounts, then, to 85.2% of 46.5%, or 39.6% of the original variation in number of hoes. Together, the two regression analyses explain 93.1% of the variation in number of hoes (53.5% in the first regression, and 39.6% in the second). Taken together, the two independent variables (site area and soil productivity) explain quite a lot of the variation in number of hoes, providing strong support for the interpretation that larger settlements had more craft workers, elites, and others not engaged in farming, and that, in addition, settlements located on more productive soils were more involved in farming. Not only are the patterns of relationships between these variables strong, they are very highly significant, which tells us that our samples, small though they may be, are large enough to give us great confidence that we are not just seeing the vagaries of sampling in operation.

Just as the assessment of the proportion of variability explained is cumulative, so are the equations for predicting the number of hoes at a site on the basis of the two independent variables. We have already produced the regression equation for predicting the number of hoes, based on the area of the site:

$$Number\ of\ hoes = (-1.959 \times Site\ area) + 47.802$$

Now we can also predict the errors in that previous estimate (that is, the residuals):

$$Residual\ number\ of\ hoes = (.010 \times Soil\ productivity) - 11.004$$

Since the residuals are the errors in the first prediction, adding the second equation to the first produces a prediction of the number of hoes at a site that is based both on its area and the productivity of the surrounding soils:

$$
\begin{aligned}
Number\ of\ hoes\ &= \{(-1.959 \times Site\ area) + 47.802\} \\
&+ \{(.010 \times Soil\ productivity) - 11.004\}
\end{aligned}
$$

There are, of course, residuals from the second regression analysis as well. If they were large enough to be interested in, we could study their relationship with yet another variable. In this way regression analysis allows for the combination of a series of analyses of relationships between two variables, and produces an integrated result of what has, in effect, become a multivariate analysis. Most statpacks will perform multiple regression, which is an extension and elaboration of this basic idea.

ASSUMPTIONS AND ROBUST METHODS

It may come as a surprise that linear regression is *not* based on the assumption that both measurements involved have normal shapes. The shape assumptions that we must be alert to in linear regression have to do with the shapes of point distributions in scatter plots. Just as we examine stem-and-leaf plots to check for the single peak and symmetry that characterize a normal shape, we examine a scatter plot prior to linear regression for the shape of point distributions. What we need to see is a cloud of points of roughly oval shape. There should be no extreme outliers from the cloud, the oval should be of similar thickness throughout, and there should be no tendencies toward curvature of the whole oval. These three potential problems can be discussed separately.

First, outliers present severe risks to linear regression. Fig. 15.9 provides an extreme example that should make the principle intuitively clear. The points in the lower left corner of the scatter plot clearly show an extremely strong negative correlation. The single outlier to the upper right, however, will cause the best-fit straight line to be as shown – a positive correlation of some strength. Outliers have such a strong effect on the best-fit straight line that they simply cannot be overlooked. When outliers are identified, those cases should be examined with great care to see whether there is a measurement or data-recording error that can be corrected or whether there is some other reason to justify excluding them from the sample.

Statpacks

Regression analysis is hardly ever performed any more except by computer. Different statpacks use a variety of vocabularies to talk about it, in part because linear regression is only the tip of the iceberg. Regression analysis is really a whole family of analytical approaches involving curved line fitting in addition to straight line fitting and incorporating a number of variables simultaneously instead of just two. Any very large and powerful statpack will perform many of these other kinds of analysis as well, and the simple, but powerful, linear regression techniques discussed here may be embedded in this broader family of analyses. Consequently, the commands or menu selections that produce a simple linear regression vary substantially from one statpack to another and are often much more complicated than it seems like they need to be. Recourse to the manual or help system for your particular program is likely to be necessary. Some statpacks integrate scatter plots into the procedures that perform regression analysis as an option, while others perform the numerical analysis as one operation and produce scatter plots as a different operation. Usually the inclusion of the curves delimiting a confidence region for the best-fit straight line is an option to be specified as part of the production of a scatterplot. Residuals, of course, are calculated as part of the regression analysis, but to be able to use them as a new measurement and pursue further analysis with them it is usually necessary to save them by specifying this as an option to the regression analysis. Typically this results in the creation of a new data file in the normal format your statpack uses for data files. The new file will have the same cases as the original data file and a variable whose values are the residuals from the regression analysis.

Second, oval shapes of points with very thin sections (or even worse, two or more separate oval clouds) are the equivalent of multipeaked shapes for single batches of numbers. They can create the same kinds of problems in linear regression that outliers do. Fig. 15.10 shows another extreme example, where two ovals of points showing negative correlations of some strength turn into a single best-fit straight line with a positive slope when improperly analyzed together. Such a shape may occur in a scatter plot of two variables that, when looked at individually, have clearly single-peaked and symmetrical shapes. Shapes like this should be broken apart for separate analysis.

Third, tendencies toward curved patterns in the oval of points can prevent a very good fit of a straight line to a fundamentally linear pattern that just happens to be curved. There are ways to extend the logic of linear regression to more complex curvilinear relationships between variables, but it is usually much easier to straighten out the curve by transforming one or both variables. The kinds of transformations required are very like the transformations discussed in Chapter 5 and may be applied to either or both of the variables to remove tendencies toward cur-

Figure 15.9. The devastating effect of a single outlier on the best-fit straight line.

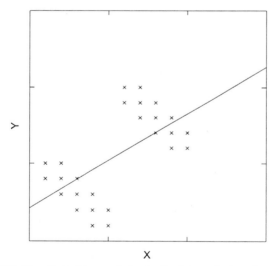

Figure 15.10. The effect of two oval clouds of points on the best-fit straight line.

vature. As Fig. 15.11 illustrates, if the scatter plot shows a tendency toward linear patterning but with the ends curving downward, a square root transformation of X will produce a straighter line. If stronger corrective action is called for, the logarithm of X can be used instead of the square root. Clearly, for the data in Fig. 15.11, the logarithm of X is too strong a transformation, having produced just as curved

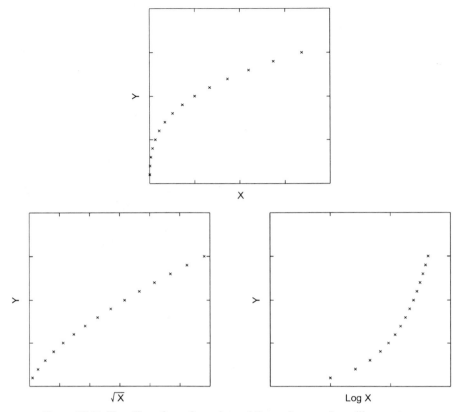

Figure 15.11. The effect of transformations of X on a downward curvilinear pattern.

a pattern in the opposite direction. Fig. 15.12 illustrates transformations to correct linear patterns where the ends curve upward. For these data the square of X produces good results. Using the cube of X produces a stronger effect than is needed in this instance. Applying, for example, a square root transformation to X prior to analysis means, of course, that it is not X but rather $\sqrt{X}$ whose relationship to Y is being investigated. Thus it becomes $\sqrt{X}$ rather than X that is used in the regression equation to predict the values of Y.

PRACTICE

You have excavated a site near Yenangyaung that has a number of apparent storage pits containing artifacts and other debris. You wish to investigate whether the density of artifacts (the number per unit volume) is constant for all the pits. (Another way

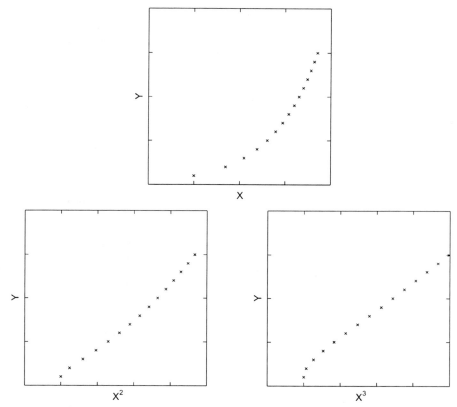

Figure 15.12. The effect of transformations of X on an upward curvilinear pattern.

to phrase this is to ask yourself whether, knowing the volume of a pit, you could accurately predict the number of artifacts it contains.) The volume measurements and the number of artifacts recovered from complete excavation of each pit are given in Table 15.4.

1. Make a scatter plot of pit volume and number of artifacts. What does inspection of the scatter plot suggest about a relationship between them?
2. Perform a regression analysis for pit volume and number of artifacts. How can the relationship between number of artifacts and pit volume be expressed mathematically? How many artifacts would you expect to find in a pit whose volume was $1.000\,m^3$?
3. How much of the variation in number of artifacts is "explained" by pit volume? What is the statistical significance of the relationship between pit volume and

Table 15.4. Data from Storage Pits at Yenangyaung

Volume (m^3)	No. of Artifacts	Volume (m^3)	No. of Artifacts
1.350	78	1.110	47
0.960	30	1.230	47
0.840	35	0.710	20
0.620	60	0.590	28
1.261	23	0.920	38
1.570	66	0.640	13
0.320	22	0.780	18
0.760	34	0.960	25
0.680	33	0.490	56
1.560	60	0.880	22

number of artifacts? Produce a scatter plot showing the 90% confidence region for the best-fit straight line.

4. Sum up clearly and concisely what this regression analysis of the relationship between pit volume and number of artifacts has shown.

Chapter 16
Relating Ranks

Sometimes we have variables that at first glance appear to be measurements, but that on further examination reveal themselves to be something less than actual measurements along a scale. Often they really amount to relative rankings rather than true measurements. For example, soil productivity is sometimes rated by producing an index with an arbitrary formula using such values as content of various nutrients, soil depth, capacity for water retention, and other variables that affect soil productivity. The formulas used in these ratings are carefully considered to produce a set of numbers such that we are sure that higher numbers represent more productive soils and lower numbers represent less productive soils. Such scales, for example, would allow us to say that a rating of 8 means more productive soils than a rating of 4. They seldom, however, leave us in position to say that a rating of 8 means soils twice as productive as a rating of 4. It is our inability to make this last statement that keeps such ratings from being true measurements. Instead, they are *rankings*. Rankings allow us to put things in rank order (most productive soil, second most productive soil, third most productive soil, etc.) but not to say *how much more* a high ranking thing is than a low ranking thing.

The logic of linear regression relies on the measurement principle. (Think of the scatter plots and the regression equations. If X is twice as large it places the corresponding point twice as far over on the scatter plot. If X is twice as large it has twice the effect on the prediction of Y by way of the regression equation.) If X is actually only a ranking rather than a true measurement, then we should feel uncomfortable about using regression. Instead of performing a linear regression and attempting to predict the actual value of Y from X, we might use a *rank order correlation coefficient* to assess the strength and significance of a rank order relationship.

A rank order relationship has nothing to do with the actual magnitude of the rankings for either variable studied, but rather only with the order of the rankings. If we rank order a batch of numbers according to the values for X and this rank order is exactly the same as the rank order of values for Y, then X and Y show a perfect

R.D. Drennan, *Statistics for Archaeologists*, Interdisciplinary Contributions
to Archaeology, DOI 10.1007/978-1-4419-0413-3_16,
© Springer Science+Business Media, LLC 2004, 2009

positive rank order relationship. That is, the highest value for X is for the case that also has the highest value for Y; the second highest value for X is for the case that also has the second highest value for Y; and so on. A perfect negative rank order relationship means that the case with the highest value for X has the lowest value for Y; the case with the second highest value for X has the second lowest value for Y; and so on until the case with the lowest value for X has the highest value for Y.

We can imagine a rank order correlation coefficient that works like Pearson's r, so that a perfect positive rank order relationship is assigned a value of 1; a perfect negative rank order relationship is assigned a value of -1; and intermediate relationships are assigned values between 1 and -1, depending on the extent to which the relationships approach one or the other of these ideal situations. Several such coefficients exist. One of the most frequently used is *Spearman's rank correlation coefficient* (r_S).

CALCULATING SPEARMAN'S RANK CORRELATION

Table 16.1 contains data for soil productivity ratings for 17 different soil zones in the Konsankoro Plain. The Neolithic occupation consisted of a series of sedentary village sites of remarkably consistent size. We take the number of village sites in

Table 16.1. Soil Productivity and Villages in the Konsankoro Plain

Soil zone	Productivity rating X	No. of villages per km^2 Y	Rankings X	Rankings Y	d	d^2	t_x	T_x	t_y	T_y
A	2	0.26	3.5	2	1.5	2.25	2	0.5	1	0.0
B	6	1.35	11.5	14	−2.5	6.25	2	0.5	1	0.0
C	3	0.44	6	6	0.0	0.00	3	2.0	1	0.0
D	7	1.26	13.5	12	1.5	2.25	2	0.5	1	0.0
E	4	0.35	8.5	4	4.5	20.25	2	0.5	1	0.0
F	8	2.30	16	17	−1.0	1.00	3	2.0	1	0.0
G	8	1.76	16	16	0.0	0.00	3	2.0	1	0.0
H	1	0.31	1.5	3	−1.5	2.25	2	0.5	1	0.0
I	3	0.37	6	5	1.0	1.00	3	2.0	1	0.0
J	5	0.78	10	11	−1.0	1.00	1	0.0	1	0.0
K	1	0.04	1.5	1	0.5	.25	2	0.5	1	0.0
L	8	1.62	16	15	1.0	1.00	3	2.0	1	0.0
M	7	1.34	13.5	13	0.5	.25	2	0.5	1	0.0
N	2	0.47	3.5	7	−3.5	12.25	2	0.5	1	0.0
O	4	0.56	8.5	9	−0.5	.25	2	0.5	1	0.0
P	3	0.48	6	8	−2.0	4.00	3	2.0	1	0.0
Q	6	0.76	11.5	10	1.5	2.25	2	0.5	1	0.0

$$\sum d^2 = 56.50 \quad \sum T_x = 17.0 \quad \sum T_y = 0.0$$

each soil zone divided by the total number of square kilometers covered by that zone to indicate how densely the zone was occupied, and we wish to investigate whether more productive soil zones were more densely inhabited.

The first step in calculating Spearman's rank correlation is to determine the rank orderings of all the cases for each of the two variables (taken separately). These rank orderings are also given in Table 16.1. Ties frequently occur in the soil productivity ratings. That is, for example, soil zones H and K are ranked in the lowest productivity category (1). These two least productive soil zones should be rank ordered 1 and 2, but we have no basis for putting one above the other since they are tied in the productivity ratings. As a consequence, we assign each a rank order of 1.5 (the mean of 1 and 2). Soil zones C, I, and P are tied with productivity ratings of 3. These would be soil zones 5, 6, and 7 in rank order if we could determine which to put above the other. Since we cannot make this determination, each is assigned a rank order of 6 (the mean of 5, 6, and 7). Such a treatment is accorded whenever there are ties. No ties occur in the number of villages per square kilometer (which actually is a true measurement), so the rank ordering is simpler. It begins at 1 for soil zone K and continues through zones A, H, and so on to zone F, which ranks 17th because it has the highest number of village sites per square kilometer.

Subtracting the rank orderings for villages per square kilometer (Y) from the rank orderings for soil productivity (X) gives us the difference between rankings d, which we then square and sum up to get $\sum d^2$.

The last four columns in Table 16.1 concern a correction that must be made for ties. The value t for each soil zone is the total number of soil zones that are tied at that ranking. For example, soil zone A has a value of $t_x = 2$ because a total of two zones (A and N) are tied at its productivity rating of 2. Since there are no ties for number of villages per square kilometer, all the values of t_y are 1. For each t value for each of the two variables, a value of T is obtained as follows:

$$T = \frac{t^3 - t}{12}$$

The calculation of Spearman's rank correlation requires three sums from Table 16.1: $\sum d^2$, $\sum T_x$, and $\sum T_y$. A sum of squares is calculated for each of the two variables:

$$\sum x^2 = \frac{n^3 - n}{12} - \sum T_x$$

where $\sum T_x$ is from Table 16.1, and n is the number in the sample (17 in this example). Thus, for the example in Table 16.1,

$$\sum x^2 = \frac{17^3 - 17}{12} - 17.0 = 408 - 17 = 391 \text{ and}$$

$$\sum x^2 = \frac{17^3 - 17}{12} - 0.0 = 408 - 0 = 408$$

Spearman's rank correlation, then, is given by the equation

Be Careful How You Say It

In conclusion to the example analysis in the text, we would say "There is a strong and highly significant rank-order correlation between soil productivity and number of villages per square kilometer ($r_S = .93$, $p < .001$)." This informs the reader that the relationship is positive (more villages in more productive soil zones), what correlation coefficient was used, and just how unlikely it is that the observed correlation would have occurred in this sample if there were no correlation in the population from which the sample was selected.

$$r_s = \frac{\sum x^2 + \sum y^2 - \sum d^2}{2\sqrt{\sum x^2 \sum y^2}}$$

For the example in Table 16.1, then,

$$r_s = \frac{391 + 408 - 56.5}{2\sqrt{(391)(408)}} = \frac{742.5}{798.8} = 0.93$$

Spearman's rank order correlation coefficient, then, between soil productivity and number of villages per square kilometer in the Konsankoro Plain is 0.93, indicating a strong positive correlation. (Values for r_S can be interpreted in much the same manner as those for Pearson's r, although the two cannot be compared directly. That is, a Spearman's r_S of 0.85 between two variables cannot be said to indicate a stronger correlation than a Pearson's r of 0.80 between two other variables.)

If there are no ties, then we can easily see that $\sum T = 0$ (as in the case of number of villages per square kilometer in Table 16.1). If there are no ties for either variable, then, there is no need to go to the trouble of figuring t and T, and the entire equation for Spearman's rank correlation is considerably simplified:

$$r_S = 1 - \frac{6 \sum d^2}{n^3 - n}$$

SIGNIFICANCE

As usual, the question of significance is "How likely is it that the correlation observed in the sample is not a consequence of a correlation in the population that the sample was selected from but instead simply a result of the vagaries of sampling?" Put another way, "How likely is it that a sample this size with a correlation this strong could be selected from a population where there is no correlation?" For samples of ten or more, this question can be answered with the familiar t table

---

Statpacks

Spearman's rank correlation coefficient is only one of several similar approaches to evaluating the strength and significance of rank order correlations. Many statpacks provide options for calculating them all under the heading of rank correlations or nonparametric correlations. Sometimes, r_S is calculated as an option with the same commands that produce Pearson's r. Even if your statpack does not provide Spearman's rank correlation as a specific option, you still may be able to trick it into producing r_S. It turns out that Spearman's rank correlation is equivalent to Pearson's r calculated on rankings. Consequently, you can provide rankings for each of your cases on the variables you are interested in (the fourth and fifth columns in Table 16.1) and use your statpack to perform a regression analysis on those variables. The resulting correlation coefficient will be equivalent to r_S.

(Table 9.1). The following formula gives the value of t:

$$t = r_S \sqrt{\frac{n-2}{1-r_S^2}}$$

In our example,

$$t = .93 \sqrt{\frac{17-2}{1-0.93^2}} = 0.93 \sqrt{\frac{15}{1-0.86}} = 0.93 \sqrt{107.14} = 9.63$$

Looking this value up in Table 9.1, using the row for $n-1 = 16$ degrees of freedom, we discover that this value of t would be far beyond the rightmost column in the table. The associated probability, then, would be far less than 0.001. Thus there is far less than one chance in 1,000 that a sample of 17 would show a Spearman's rank correlation this strong if it had been selected from a population where there was no rank order relationship between the two variables.

It should be noted that this example raises some complicated questions of what population the data are a sample from. The sample consists of 17 soil zones that have been surveyed. In order to accomplish the analysis we have just done, we must take these 17 soil zones as a random sample from a larger and vaguely defined population of soil zones that are or might be in the Konsankoro Plain. This sample has given us what we take to be 17 separate and independent observations for the two variables, and these 17 observations form the batch that we have analyzed as a sample. Strictly speaking, this is not a random sample from a population of soil zones. Indeed, this sample may represent a complete survey of the entire Konsankoro Plain. If we have studied the entire population, it may seem to make little sense to treat the data as a sample. In evaluating significance, however, we frequently engage in a sort of

Table 16.2. Probability Values for Spearman's Rank Correlation r_s
for Samples of Less Than 10[a]

Confidence	80%	90%	95%	99%
	.80	.90	.95	.99
Significance	20%	10%	5%	1%
	.20	.10	.05	.01
n				
4	.639	.907	1.000	
5	.550	.734	.900	1.000
6	.449	.638	.829	.943
7	.390	.570	.714	.893
8	.352	.516	.643	.833
9	.324	.477	.600	.783

[a] (Adapted from "Distributions of Sums of Squares of Rank Differences for Small Numbers of Individuals" by E.G. Olds *(Annals of Mathematical Statistics* 9:133–148 [1938])

pretend sampling from an imaginary larger population. What we learn from the evaluation of significance in a case like this is still, however, whether we should have much confidence in the correlation observed. What we have found out in this instance is that the correlation we observed is not at all likely to be pure random chance at work in a small sample. We will consider this notion of pretend sampling further in Chapter 20.

The formula for values of t is appropriate only if the sample is ten or more. If the size of the sample is less than ten, then Table 16.2 should be used to determine the associated probability.

ASSUMPTIONS AND ROBUST METHODS

Since Spearman's rank correlation does not assume normal distributions, or rely on means, standard deviations, or scatter plots, it is automatically highly robust. No transformations or other modifications need ever be applied. This, in effect, makes r_S a very robust correlation coefficient that can be used instead of Pearson's r when such factors present problems for the application of Pearson's r.

PRACTICE

You have excavated the remains of 12 dwellings in the village site of Teixeira. You notice that some of the artifacts recovered from the dwelling areas are finer and

**Table 16.3. Floor Area and Artifact Status Index for
12 Excavated Houses from the Teixeira Site**

Status index	Floor area (m^2)
23.4	31.2
15.8	28.6
18.3	27.3
12.2	22.0
29.9	45.3
27.4	33.2
24.2	30.5
15.6	26.4
20.1	29.5
12.2	23.1
18.5	26.4
17.0	23.7

fancier than others, and might indicate differences in status or wealth between the households. You identify a variety of ornamental objects and pottery with incised decoration as possible status indicators, and you count the number of such artifacts in each household area per 100 artifacts recovered. This gives you an index of status or wealth based on the artifact assemblages in the different households. You wish to investigate whether this status index is related to the size of the dwelling structure itself (pursuing the idea that wealthier families might have larger houses). The data are given in Table 16.3

1. How strong and how significant is the relationship between house floor area and your status index?
2. What sort of support do your observations provide for the idea that wealthier households (as indicated by their possessions) had larger houses?

Part IV
Special Topics in Sampling

Chapter 17
Sampling a Population with Subgroups

When the population we are interested in has subgroups that we are also interested in separately, it is often useful to select a separate sample of elements from each of the subgroups. For such purposes each subgroup is treated as if it were a completely separate population. A sample of whatever size is needed is selected from each of these separate populations, and the values of interest are estimated separately for each population. Suppose that we have reliable information on the locations of all sites in a region. No one has attempted to discover the sizes of these sites, however. We could select a sample of the known sites and go make systematic surface collections in an effort to determine how large they are. These determinations could then form the basis for estimating the mean site size for the region. If, in addition, the region could be divided into three different environmental settings (remnant levees, river bottoms, and slopes) we might be interested in estimating the mean site area for each of the settings.

Table 17.1 provides information on a sample of sites for each of these three settings, as well as a stem-and-leaf plot for each sample. The table gives N, the total number of sites in each setting (the three populations sampled), and n, the number of sites in each of the three samples. The stem-and-leaf plots show a single-peaked and symmetrical shape for each of the samples, and their standard errors have been calculated using the finite population corrector (Chapter 9) since the sampling fractions are large. Multiplying these standard errors by the corresponding values of t for 95% confidence and $n - 1$ degrees of freedom gives us error ranges to attach to the estimated mean site areas for each of the three settings. Thus we are 95% confident that the mean area of sites on remnant levees is 1.71 ha $\pm$ 0.32 ha; in the river bottoms, 2.78 ha $\pm$ 0.31 ha; and on the slopes, 0.83 ha $\pm$ 0.32 ha.

R.D. Drennan, *Statistics for Archaeologists*, Interdisciplinary Contributions to Archaeology, DOI 10.1007/978-1-4419-0413-3_17, © Springer Science+Business Media, LLC 2004, 2009

Table 17.1. Site Areas (ha) in Three Settings

River Bottoms			Remnant Levees			Slopes		
$N = 53$			$N = 76$			$N = 21$		
$n = 12$			$n = 19$			$n = 7$		
$\overline{X} = 2.78$			$\overline{X} = 1.71$			$\overline{X} = 0.83$		
$SE = 0.14$			$SE = 0.15$			$SE = 0.13$		
3.3			2.9			0.7		
2.7	4		1.7	4		1.3	4	
2.1	3	8	1.3	3		1.2	3	
3.8	3	134	2.1	3	2	0.6	3	
2.7	2	7789	1.9	2	59	0.6	2	
3.4	2	144	1.2	2	0113	1.2	2	
2.9	1	8	2.5	1	66779	0.2	1	
2.8	1		2.1	1	0234		1	223
2.4	0		1.6	0	78		0	667
1.8	0		1.7	0	4		0	2
2.4			2.0					
3.1			1.6					
			1.0					
			1.4					
			2.3					
			3.2					
			0.8					
			0.4					
			0.7					

These estimates and their 95% confidence error ranges confirm what we might well have suspected from looking at the three stem-and-leaf plots – sites in the three settings have markedly different mean sizes, and the differences that we observe between our three samples are not at all likely to be just the result of sampling vagaries. Up to this point, we have done nothing more than treat these three samples in the ways discussed in Chapter 9.

POOLING ESTIMATES

At this point, however, we might well want to consider the three samples together in order to talk about sites in the region in general, irrespective of the settings in which they were located. We cannot simply put all the sites from all three samples together into one sample, though, and consider it a random sample of sites in the region. Such a sample would most definitely not be a random sample of the sites in the region because the selection procedures did not give each site in the region an equal chance of selection. Of the 21 sites on the slopes, 7 (or 33.3%) were selected;

of the 53 sites in the river bottoms, 12 (or 22.6%) were selected; and of the 76 sites on remnant levees, 19 (or 25.0%) were selected. Thus river bottom sites had less chance of being included in the sample (a probability of 0.226) than sites on levees (a probability of 0.250), and levee sites had less chance of being included than sites on the slopes (a probability of 0.333). The overall sample produced by just putting these three separate samples together would systematically over-represent slope sites and systematically under-represent river bottom sites. Any conclusions we might arrive at about mean site area in the region as a whole based on such a sample would be affected by these sampling biases.

What we must do is consider the larger problem one of *stratified sampling*, as selecting separate samples from different subgroups of a population is usually called. In this example, each of the three environmental settings would be a *sampling stratum*. Each sampling stratum would form a population to be sampled separately from the other sampling strata, just as we have done in this example. Appropriate sample sizes and sampling procedures would be determined independently for each sampling stratum, and the samples selected would be used independently to make estimates about each of the parent populations. We have already done all of this. It raises no new issues in sampling beyond those dealt with in Chapters 7–11.

Only at the last step, that of *pooling* the estimates made for each sampling stratum into an overall estimate for the whole population must special steps be taken. In the first place, having already discovered that sites in the three different settings have rather different mean areas, we must consider whether it makes any sense even to speak of the mean area of sites for the region as a whole. If the overall population of sites had a shape with multiple peaks, it would be foolish to attempt any analysis of the entire set of sites as a single batch. We do not, of course, have any way of knowing for certain what the shape of the whole population would be, but, since the sampling fractions in the three sampling strata are not wildly different, we could look at a stem-and-leaf plot of all three samples together to get a rough idea. Such a stem-and-leaf plot appears in Table 17.2. It is certainly single peaked and symmetrical enough to make it meaningful to use the mean as an index of center for the whole batch. Thus, we could consider it sensible to make an estimate of

**Table 17.2. Stem-and-Leaf Plot
of Areas of Sites from
All Three Samples in Table 17.1**

4	
3	8
3	1234
2	577899
2	0111344
1	667789
1	0222334
0	66778
0	24

the mean site area for all sites in the region by pooling the estimates for the three sampling strata, as follows:

$$\overline{X}_p = \frac{\Sigma\left(N_h\overline{X}_h\right)}{N}$$

where $\overline{X}_p$ = the pooled estimate of the mean, that is, the estimated mean for the entire population, taking all sampling strata together, $\overline{X}_h$ = the mean of the elements in the sample for stratum h, N_h = the total number of elements in the population of stratum h, and N = the total number of elements in the entire population.

For the example from Table 17.1,

$$\overline{X}_p = \frac{(76)(1.71)+(53)(2.78)+(21)(.83)}{150} = \frac{294.73}{150} = 1.96\,\text{ha}$$

Thus we estimate that the mean area of sites in the region as a whole (irrespective of environmental setting) is 1.96 ha. We attach an error range to this estimate in a similar fashion, by pooling the standard errors for the three separately selected samples:

$$SE_p = \frac{\sqrt{\Sigma\left(N_h^2\right)\left(SE_h^2\right)}}{N}$$

where SE_p = the pooled standard error for all sampling strata taken together, SE_h = the standard error for sampling stratum h, N_h = the total number of elements in the population of stratum h (as before), and N = the total number of elements in the entire population (also as before).

For the example from Table 17.1,

$$SE_p = \frac{\sqrt{(76^2)(.15^2)+(53^2)(.14^2)+(21^2)(.13^2)}}{150} = \frac{13.87}{150} = .09$$

This pooled standard error is treated like any other. To produce an error range for 95% confidence, we would multiply it by the value of t corresponding to 95% confidence and $n-1$ degrees of freedom where n is now the number in all three samples considered together, or 38. This value of t is 2.021, so we would be 95% confident that the mean area of all sites in the region is 1.96 ha $\pm$ 0.18 ha.

THE BENEFITS OF STRATIFIED SAMPLING

Stratified sampling can sometimes offer a more precise estimate for an entire population than simply sampling the entire population directly. This makes stratified sampling potentially useful even in situations where we might not be much interested in the separate means of the sampling strata. The possible increased precision comes from providing a smaller error range in the situation where a population has subgroups whose means differ somewhat from each other but which have very small standard deviations when each is taken separately. That is, if the subgroups each form batches with smaller spreads than the population as a whole, the error ranges associated with the estimates of their means may be quite small. When these are pooled into an error range for the estimated overall population mean it may well be smaller than the error range that would have been obtained from a single sample drawn randomly from the population as a whole. Sometimes this effect is strong enough to outweigh the opposite effect resulting from the fact that the samples from the subgroups are each smaller than the total sample. If a population is easily divided into subgroups whose means may be different and whose members vary little from each other, then it is worth considering sampling that population by those subgroups instead of as a whole, even if the subgroups are of little intrinsic interest separately.

Chapter 18
Sampling a Site or Region with Spatial Units

Sometimes the sampling elements available for selection are not the same as the elements we wish to study. This happens most frequently in archaeology in *spatially based sampling*, as in the excavation of a sample of grid squares in a site or the survey of a sample of grid squares or transects in a region. For instance, suppose we have a random sample of 500 sherds from a site. We may want to estimate, say, the mean thickness of sherds at the site or the percentage of a particular pottery type in the sherds at the site. The elements studied are sherds. Suppose the sample had been obtained by excavating a random sample of ten grid squares. The sampling element here is not the sherd but the grid square. It was ten grid squares that were randomly selected from all the squares in the site grid, not 500 sherds from all the sherds in the site. We thus have a sample, not of 500 independently selected elements, but of ten independently selected elements, and these elements do not correspond to the elements we need to study. Each sampling element is, in this case, a group or *cluster* of a varying number of the elements of study (sherds). This fact must be allowed for in making estimates of means or proportions.

Estimating population means and proportions from samples, and attaching error ranges to those estimates was the subject of Chapters 9 and 11. This chapter extends that discussion to the special case where the sampling elements are different from the elements of study. This chapter on *cluster sampling*, then, can be considered a special case of the general topics dealt with in Chapters 9 and 11, which can be referred to as *simple random sampling* to distinguish them from more complex kinds of sampling. Cluster sampling is particularly important in archaeology because so much of the sampling we do is based on spatial units.

R.D. Drennan, *Statistics for Archaeologists*, Interdisciplinary Contributions
to Archaeology, DOI 10.1007/978-1-4419-0413-3_18,
© Springer Science+Business Media, LLC 2004, 2009

SPATIAL SAMPLING UNITS: POINTS, TRANSECTS, AND QUADRATS

At least three different kinds of spatial sampling units might be used in archaeology – *points*, *transects*, and *quadrats*. True transects are almost never used in archaeology, but the occasion for point sampling does sometimes arise. An example might be in a region subject to substantial alluviation that has buried archaeological sites beneath thick layers of sediment. An effort to estimate, say, the total area of sites in the region might pursue a sampling program based on drilling cores down into the sediments. Such cores could be large enough in diameter to recover recognizable artifacts if habitation deposits were intersected, making it possible to say that a particular core was either within a site or not within a site. The cores might well be treated as point observations: either site or not site. If a series of random locations was selected for coring, these observations could be treated as a random sample of the total area of the region. The proportion of cores that were within sites would be an estimate of the proportion of the region's total area that is within sites. That is, if the region covered $100 \, \text{km}^2$, and if 5% of the cores produced artifacts, we would estimate that 5% of the region's $100 \, \text{km}^2$ was within sites. Thus our estimate of the total area of the sites in the region would be $5 \, \text{km}^2$. Since in this case, the units of sampling were points in space and the estimated proportion is a characteristic of the space itself, it is not an instance of cluster sampling but of simple random sampling, and an error range could be attached to this estimate following the procedure discussed in Chapter 11. It would depend on the size of the sample (n), which would be the number of cores drilled. The population (N) would be infinite. The practice questions at the end of Chapter 6 actually comprise an example of point sampling.

Quadrats (not *quadrants*, which are something else) are two-dimensional spatial units, in archaeology most often the squares in a grid system. They can also be rectangles or other shapes. When the rectangles are long and narrow and run from one side of the study area to the other, we often refer to them as *transects*, but technically these are *quadrats*. True transects, like lines, have only length; their width is 0. When an archaeologist walks along a "transect" from one side of a survey zone to the other, the observations are not actually along a line but within a very long narrow rectangle including some distance to either side of the path walked. Such "transects" are usually best treated as long narrow quadrats in cluster sampling since they do have a width (and thus an area) based on how far to either side the archaeologist can observe whatever is to be observed.

Perhaps the most frequently used method of selecting a random sample of quadrats is to lay out a grid dividing the area to be sampled (say, a site to be excavated) into sampling units. Each potential excavation unit in the grid system can be assigned a number beginning with 1, and a random number table can be used to select a sample of these quadrats. One possible result of such a sampling scheme is shown in Fig. 18.1. The same system can be used for long narrow quadrats ("transects"). In this case the grid divides the area to be sampled into long narrow rectangles running from one side to the other, each as wide as the coverage of a single "transect." These are assigned numbers for random selection (Fig. 18.2).

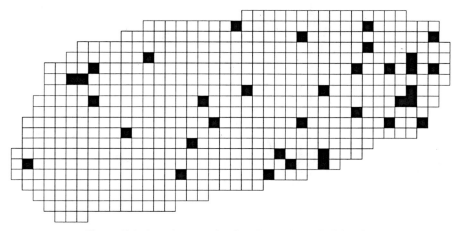

Figure 18.1. A random sample of quadrats selected individually.

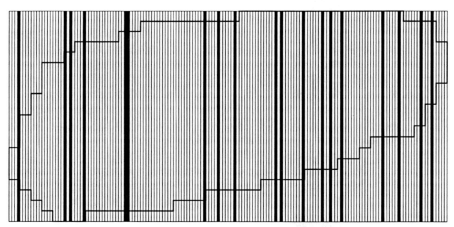

Figure 18.2. A random sample of "transects" (actually very long narrow quadrats) selected individually.

Another interesting possibility, sometimes used to avoid having all "transects" parallel to each other, is to enclose the area to be sampled in a rectangular frame on a map and place tick marks on all sides of the frame. The ticks should be as far apart as the "transects" are wide. The ticks are numbered sequentially, beginning at any point with 1 and continuing all the way around the frame until the starting point is reached again. A random number determines one end of the first "transect," and a second random number determines its other end. (If the second random number indicates a tick mark on the same side of the frame as the first, it is discarded and another is selected.) The process is repeated until the desired number of "transects" has been selected. One result of such a sampling scheme is illustrated in Fig. 18.3.

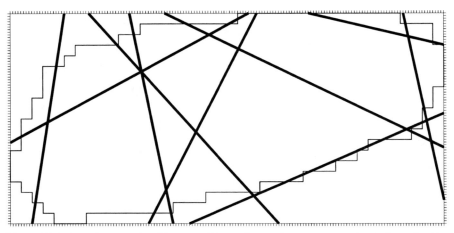

Figure 18.3. A random sample of "transects" (actually very long narrow quadrats) determined by random selection of their endpoints.

When n random quadrats are selected from all the quadrats in the grid at large, it is often the case that some sample quadrats are very close to each other (possibly even adjacent), and one or more fairly large parts of the study area may be left entirely unsampled. Figure 18.1 shows both these characteristics. This may be unsatisfactory, for example, in excavating a site by random sampling, since (for reasons not related to random sampling) we might not want to leave one whole section untested. One alternative sometimes applied in such situations is *systematic sampling*. As an example, suppose that we want a sample of 36 quadrats from an area consisting of 570 quadrats. To select a systematic sample, we would subdivide the grid of 570 quadrats into 36 subsets consisting of 16 contiguous quadrats each. (We would add six dummy quadrats, indicated with hatching in Fig. 18.4, to fill out the full 16 in each subset.) One quadrat would be randomly selected from each subset of 16 by repeatedly selecting random numbers between 1 and 16. The sample might be as large as 36, but if any dummy quadrats are selected, then the final sample is less than 36. The dummy quadrats never really become part of the sample, even if they are selected, because they are imaginary. Their function is to provide each real quadrat exactly one chance in 16 of being selected for the sample. The resulting sample could still include adjacent quadrats, as Fig. 18.4 shows, but the large unsampled areas that frequently occur in simple random samples of quadrats would be impossible.

It is sometimes objected that systematic sampling is not strictly random, and technically this is true. In strict terms, the selection of one element in a random sample should not have any effect at all on the selection of other elements. The selection of a quadrat in a systematic sample, however, causes the other quadrats in the same subset to lose their eligibility for future selection. Perhaps more important, systematic sample selection, as described here, comprises sampling without replacement since a quadrat, once selected, is no longer available for future selection. The equations

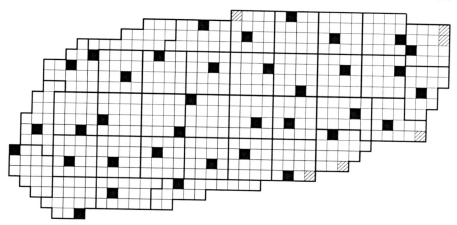

Figure 18.4. A systematic random sample of quadrats.

given in this chapter, like those in Chapters 9 and 11, are for sampling with replacement. The impact of these two technical problems, however, is minimal under most circumstances. (And especially for the form of systematic sampling described here, which violates the strictest norms of random sampling less than other variants of systematic sampling that have been suggested.) The attractiveness of working with a spatial sample that avoids leaving any large sections of the study area unexamined usually outweighs these minor technical objections.

ESTIMATING POPULATION PROPORTIONS

The best estimate of a population proportion in cluster sampling is the same as in the case of simple random sampling – simply the proportion in the sample. For instance, suppose we excavate a random sample of ten grid squares in a site and obtain 500 sherds altogether. If 35% of this cluster sample of sherds are cord marked, we would estimate that 35% of the sherds in the site are cord marked.

The error range corresponding to this estimate, as with simple random sampling, is based on the standard error of the proportion, but the standard error of the proportion in cluster sampling is calculated by the formula

$$SE = \sqrt{\left(\frac{1}{n}\right)\left(\frac{\sum\left(\frac{x}{y}-P\right)^2 \left(\frac{yn}{Y}\right)^2}{n-1}\right)\left(1-\frac{n}{N}\right)}$$

where SE = standard error of the proportion, n = sample size (i.e., number of units in the sample), N = population size (i.e., number of units in the population), x = number of object x in a unit, y = number of object y in a unit, P = estimate of the

**Table 18.1. Sherds from a Random Sample of
Ten Excavation Units**

Unit	x No. of cord-marked sherds	y No. of sherds
07	10	32
18	13	27
29	16	38
31	19	73
37	17	55
56	21	41
72	18	63
83	30	81
87	19	56
91	12	34

proportion x/y for the population, and $Y = \sum y$, that is, the total number of object y in n units.

Note that this formula incorporates the finite population corrector $1 - (n/N)$ discussed in Chapter 9. Since the population is usually finite and definable in spatial sampling, this correction can usually be applied. If the population is very large, however, compared to the size of the sample, the finite population corrector has a negligible effect on the result because it is only trivially different from one.

Table 18.1 provides an example of the calculations involved. It describes a random sample of ten excavated grid units from a site whose total area is 100 grid units. The sample size, n, is thus 10, and the population size, N, is 100. These ten excavation units yielded a total of 500 sherds, some of which were cord marked. We wish to estimate the proportion of cord-marked sherds in the ceramic assemblage of the site as a whole. The units are identified by their sequential numbers, which were used to select the random sample. Since the proportion we wish to estimate is the proportion of cord-marked sherds in the ceramic assemblage, x is the number of cord-marked sherds in each square and y is the total number of sherds in each square. $\sum x$, or X, is 175 – the total number of cord-marked sherds found in all ten units. And $\sum y$, or Y, is 500 – the total number of sherds found in all ten units. The proportion of cord-marked sherds in the ceramic assemblage for the entire sample, then, is $(175/500) = 0.350$; 35.0% of the sherds are cord marked. Since the best estimate of the population proportion is the sample proportion, we would estimate that 35.0% of the sherds in the site are cord marked.

Table 18.2 extends Table 18.1 into a step-by-step calculation of the summation needed for the standard error calculation. The first calculation step (x/y) results in the fourth column, simply dividing x by y for each excavation unit. This quantity is, of course, the proportion of cord-marked sherds for each of the ten excavated units. This proportion varies from a low of 26.0% in unit 31 to a high of 51.2% in unit 56. It is this variation from one sample unit to the next in the proportion of interest to us that will form the basis for the error range.

Table 18.2. Calculation of the Summed Weighted Deviations from the Overall Sample Proportion

Unit	x	Y	$\dfrac{x}{y}$	$\dfrac{x}{y}-P$	$\left(\dfrac{x}{y}-P\right)^2$	$\dfrac{yn}{Y}$	$\left(\dfrac{yn}{Y}\right)^2$	$\left(\dfrac{x}{y}-P\right)^2\left(\dfrac{yn}{Y}\right)^2$
07	10	32	0.313	−0.037	0.001369	0.640	0.409600	0.000561
18	13	27	0.481	0.131	0.017161	0.540	0.291600	0.005004
29	16	38	0.421	0.071	0.005041	0.760	0.577600	0.002912
31	19	73	0.260	−0.090	0.008100	1.460	2.131600	0.017265
37	17	55	0.309	−0.041	0.001681	1.100	1.210000	0.002034
56	21	41	0.512	0.162	0.026244	0.820	0.672400	0.017646
72	18	63	0.286	−0.064	0.004096	1.260	1.587600	0.006503
83	30	81	0.370	0.020	0.000400	1.620	2.624400	0.001050
87	19	56	0.339	−0.011	0.000121	1.120	1.254400	0.000152
91	12	34	0.353	0.003	0.000009	0.680	0.462400	0.000004

$$\sum y = 500 \qquad\qquad\qquad \sum \left(\frac{x}{y}-P\right)^2\left(\frac{yn}{Y}\right)^2 = 0.053132$$

For the fifth column in the table, $x/y - P$, the overall sample proportion (0.350) is subtracted from each excavation unit's proportion. This is recognizable, of course, as a deviation – the extent to which the proportion of cord-marked sherds in each excavation unit deviates from the overall proportion in the whole sample taken together. In a manner familiar from all our calculations of standard deviations, the next step, $(x/y - P)^2$, squares the deviations from the fifth column to arrive at the sixth column. The sixth column is one of the two terms that must be multiplied together to arrive at the quantity to be summed.

The other term is, in effect, a weighting factor. The sixth column is a set of squared deviations. In cluster sampling we weigh more heavily the deviations of units that produce more evidence (that is, in this example, more sherds). This makes some intuitive good sense if you remember that generally we get a more accurate estimate from a larger sample than from a smaller sample. In effect, each excavation unit is a single sample of sherds from the site. These single samples can be expected to produce somewhat different results – that is, to deviate from the overall proportion. If they all deviate very little from the overall proportion, then the error range associated with our cluster sampling estimate should be relatively small. (The consistency from one excavation unit to another makes us willing to believe the site is fairly homogeneous and our estimate fairly precise.) We will be more concerned about imprecision in our results if units that produce large numbers of sherds deviate widely from the overall proportion than if units producing small numbers of sherds deviate widely from the overall proportion. Thus, when we sum up the squared deviations, we will count the deviations in units that produced large numbers of sherds more heavily than deviations in units that produced small numbers of sherds (where the deviations are more likely to be simply the result of random vagaries in smaller samples).

The seventh column begins the calculation of this weighting factor, which is based on how "large" the sample unit is in terms of the elements we are studying. In this example the elements we are studying are sherds, so the large (heavily weighted) sample units are those that produced large numbers of sherds. The seventh column (yn/Y) is simply the number of sherds in a unit times the number of units in the sample (10) divided by the total number of sherds in the sample (500). (It is useful to know, for checking calculations, that the sum of the seventh column is always n, the number of units in the sample.) The eighth column is simply the square of the seventh column.

The last column in Table 18.2 is the product of the sixth column and the eighth column, and is the quantity to be summed for all ten sample units: $(x/y-P)^2(yn/Y)^2$. The sum of this quantity for all sample units is indicated at the bottom of the last column: 0.053132.

Substitution of numbers into the formula given above for the standard error of the proportion is now relatively straightforward:

$$SE = \sqrt{\left(\frac{1}{10}\right)\left(\frac{.053132}{10-1}\right)\left(1-\frac{10}{100}\right)}$$
$$= \sqrt{(.100)(.005904)(.900)}$$
$$= \sqrt{.00531} = .023$$

The standard error of the proportion, then, is 0.023, so the estimate of the proportion of cord-marked sherds in the ceramic assemblage at the site could take the form $35.0\% \pm 2.3\%$. This error range can be increased by multiplying it by an appropriate value of t to make it a statement at whatever level of confidence is desired (see

---

Statpacks

Very few computer statpacks provide for the calculation of standard errors for cluster samples. They can certainly be calculated by hand, although the summation that forms the numerator of the middle fraction is tedious. The ease with which this calculation is illustrated as a table in which each column is derived by relatively simple repeated calculation from the previous one, however, suggests a computerized solution. Spreadsheets were designed precisely for performing such calculations, and are probably the fastest, easiest, and most commonly available option for getting a computer to do most of the boring work. The provisions that statpacks provide for transforming variables can also often be adapted to this task, since each column in Tables 18.2 and 18.4 really is simply a new variable whose values are calculated by a repeated mathematical manipulation from previous columns. Some database managers also provide mathematical tools that can perform calculations like this.

Chapter 9). For example, if we wish to express our estimate with an error range at the 95% confidence level, we look in Table 9.1 for the value of t corresponding to 95% confidence and nine degrees of freedom $(n - 1)$. This value is 2.262. Thus the error range we seek is $(2.262)(0.023) = 0.052$. We would thus say that we are 95% confident that cord-marked sherds represent $35.0\% \pm 5.2\%$ of the sherds at this site.

ESTIMATING POPULATION MEANS

As with proportions, the best estimate of the population mean is the overall mean in the sample. Table 18.3 provides example data from the same sample of excavated units we have been considering. The data here are lengths in millimeters of the projectile points encountered in these same excavation units. Altogether 21 projectile points were recovered, with an overall mean length of 21.8 mm. We would thus estimate that the mean length of all the projectile points in the site is 21.8 mm.

The standard error we need in order to put an error range with this estimated mean is calculated in a manner very similar to the standard error of the proportion:

$$SE = \sqrt{\left(\frac{1}{n}\right)\left(\frac{\sum\left(\bar{x} - \overline{X}\right)^2 \left(\frac{yn}{Y}\right)^2}{n - 1}\right)\left(1 - \frac{n}{N}\right)}$$

where $SE =$ standard error of the mean, $n =$ sample size (i.e., number of clusters in the sample), $N =$ population size (i.e., number of clusters in the population), $\bar{x} =$ mean of x in a cluster, $\overline{X} =$ estimated population mean of x (i.e., the overall sample mean), $y =$ number of things in a cluster measured for x, and $Y = \sum y$ (i.e., the total of y in n clusters).

Table 18.3. Lengths of Projectile Points from a Sample of Ten Excavation Units

Unit	X Projectile point lengths (mm)
07	15, 19, 23
18	17
29	18, 23
31	18, 18, 27
37	18, 19
56	24
72	20, 21, 26, 28, 29
83	16
87	28
91	25, 26

Table 18.4. Calculation of Summed Deviations from Overall Sample Mean

Unit	Y	$\bar{x}$	$\bar{x}-\bar{X}$	$(\bar{x}-\bar{X})^2$	$\dfrac{yn}{Y}$	$\left(\dfrac{yn}{Y}\right)^2$	$(\bar{x}-\bar{X})^2\left(\dfrac{yn}{Y}\right)^2$
07	3	19.0	−2.8	7.840	1.429	2.042	16.009
18	1	17.0	−4.8	23.040	0.476	0.227	5.230
29	2	20.5	−1.3	1.690	0.952	0.906	1.531
31	3	21.0	−0.8	0.640	1.429	2.042	1.307
37	2	18.5	−3.3	10.890	0.952	0.906	9.866
56	1	24.0	2.2	4.840	0.476	0.227	1.099
72	5	24.8	3.0	9.000	2.380	5.664	50.976
83	1	16.0	−5.8	33.640	0.476	0.227	7.636
87	1	28.0	6.2	38.440	0.476	0.227	8.726
91	2	25.5	3.7	13.690	0.952	0.906	12.403

$\bar{X}=21.8$ $\qquad\qquad\qquad\qquad\qquad\qquad \sum(\bar{x}-\bar{X})^2\left(\dfrac{yn}{Y}\right)^2 = 114.783$

Once again, the complicated calculation is the summation that forms the numerator of the middle fraction, and this summation is quite similar to the summation required for the standard error of the proportion. Table 18.4 shows this calculation carried out.

In this instance, y is the number of projectile points found in each unit. For each excavation unit, we calculate a mean projectile point length based on the projectile points in that unit. This mean appears in the third column of Table 18.4. The deviation we are interested in this time is the difference between the mean projectile point length for each unit and the overall mean in the sample (the fourth column). As usual, this deviation is squared (the fifth column). The weighting factor works just as it did in the case of estimating proportions (the sixth and seventh columns). The final product is summed in the last column.

The results can then be substituted in the formula on the preceding page as follows:

$$SE = \sqrt{\left(\frac{1}{10}\right)\left(\frac{114.783}{10-1}\right)\left(1-\frac{10}{100}\right)}$$
$$= \sqrt{(.100)(12.7537)(.900)}$$
$$= \sqrt{1.1478} = 1.07\,\text{mm}$$

To use this standard error as an error range at the 95% confidence level, as we did with the standard error of the proportion earlier in the example, we would multiply it by the value of t for nine degrees of freedom and 95% confidence, which continues to be 2.262. Thus the error range would be $(1.07)(2.262) = 2.42\,\text{mm}$, and we would be 95% confident that the mean length of all the projectile points in the site is $21.8\,\text{mm} \pm 2.4\,\text{mm}$.

DENSITIES

The fact that spatially based sampling often takes us into the realm of cluster sampling rather than simple random sampling should not confuse us about what the basic principle of cluster sampling is. Sometimes spatially based sampling is actually simple random sampling. It depends on what the elements to be studied are. In the two examples discussed above the elements to be studied were sherds and projectile points. Neither of these elements was the same as the sampling unit, since quadrats from a grid were the things randomly chosen to define the sample. Thus it was necessary to consider both estimating proportions of cord-marked sherds and estimating mean projectile point length as questions in cluster sampling.

Often, however, we are interested in studying the very spatial units that are randomly selected to form the sample. Something like this may be the case with point sampling, as noted at the beginning of this chapter. It also can happen when we estimate the density (the number per unit area) of some artifact or feature. Such densities are usually easily expressed as numbers of things, say sherds, per grid unit. Such numbers are actually properties, not of the sherds, but rather of the grid units. The elements being studied are the same as the elements that were randomly selected to form the sample, so this becomes a question in simple random sampling. In the example above, we had a sample of ten grid units that produced, respectively, 32, 27, 38, 73, 55, 41, 63, 81, 56, and 34 sherds. We can estimate the mean number of sherds per grid unit in the complete population of grid units (that is, the entire area of the site) in exactly the manner discussed in Chapter 9 for a sample of ten with a measurement for each of the ten units.

The mean number of sherds per grid unit in the sample is 50.0. Thus we would estimate that in the site as a whole the mean number of sherds per grid unit is 50.0. The standard error is 5.8 sherds, which the finite population correction reduces to 5.5 sherds, so an error range at the 95% confidence level would be $(2.262)(5.5) = 12.4$ sherds. We are thus 95% confident that the mean number of sherds per grid unit in the site is 50.0 ± 12.4 sherds.

Such estimates of densities are the most direct springboard to estimates of the total quantities of various things in the site. For example, we have estimated that there is an average of 50.0 ± 12.4 sherds per grid unit in the site (at a 95% confidence level). We know that the entire site consists of 100 grid units. Our density estimate thus translates into an estimate that the total number of sherds in the site is (50.0 sherds per grid unit) (100 grid units) = 5,000 sherds. The error range also translates in the same way: (12.4 sherds per grid unit) (100 grid units) = 1,240 sherds. We are thus 95% confident that the total number of sherds in the site is $5,000 \pm 1,240$.

Chapter 19
Sampling without Finding Anything

Sampling statistics ordinarily take as their point of departure some finding in a sample. Say the sample consists of artifacts, including some projectile points. We can estimate the proportion of projectile points in the artifact assemblage from which the sample came; we can estimate the mean weight of projectile points for the population of projectile points from which the sample came; we can estimate the proportions of different raw materials of which projectile points in the population were made; and so on. Following the procedures discussed in Chapters 9, 11, and 18, we can attach error ranges for particular confidence levels to these estimates.

Sometimes, however, we have particular reason to be interested in some specific category of observation that just does not appear at all in a sample. For example, we may recognize chert, flint, and obsidian as potentially available raw materials from which projectile points could be made, but perhaps our sample includes only chert and flint points. How confidently can we say that obsidian was not used to make projectile points? We certainly know enough about samples by now to know that the fact that we find no obsidian projectile points in a sample does not necessarily mean that there were none at all in the population from which the sample was drawn. This is true no matter how large the sample is. The only way to be certain that there are no obsidian projectile points is to acquire and study the entire population of projectile points. As long as one projectile point remains unexamined, there is at least some possibility that it could be made of obsidian.

As long as we are working with a sample, then, we must settle for some level of confidence short of 100%, just as in all the conclusions we have made about populations on the basis of samples. The confidence we have, as always, will depend on the size of the sample. We will be more confident in saying that projectile points were not made of obsidian if we have failed to find any obsidian points in a sample of 100 points than if we have failed to find them in a sample of five points. The proportion of obsidian projectile points in the population is also involved. It is intuitively obvious that if the population really includes many obsidian projectile points, it is more likely that at least one will turn up in a sample of a given size than if there are very few obsidian points in the population.

The aim of this chapter is to put a finer point on these intuitive (but perfectly valid) approximations. Applying basic statistical principles to the task requires only deciding at what level of confidence we need to speak and how many obsidian

R.D. Drennan, *Statistics for Archaeologists*, Interdisciplinary Contributions
to Archaeology, DOI 10.1007/978-1-4419-0413-3_19,

projectile points we are willing to risk overlooking. Deciding at what level of confidence to speak presents no novel aspect in this context; all the considerations brought to bear in previous chapters apply. Deciding how many projectile points we are willing to risk overlooking, however, does raise a new issue, and making this decision is what enables us to put statistical tools to good use here. In effect, we must decide what low proportion of obsidian projectile points is functionally equivalent to none. If only one projectile point in a billion were made of obsidian, we would presumably be willing (for many purposes at least) to say that, in effect, obsidian was not used for projectile points. We would probably be equally willing to say that one obsidian point in a million really meant that points were not made of obsidian. For some purposes at least, it would be interesting and useful to be able to say with high confidence that fewer than 1% or even 5% of the projectile points in some population were made of obsidian.

Suppose that we have a sample of 16 projectile points, and none is made of obsidian. We would like to know at what level of confidence we can say that fewer than 1% of the projectile points in the population from which the sample was selected were made of obsidian. Another (and more familiar) way to put this question is, "How likely is it that we could select a random sample of 16 including no obsidian projectile points from a population with as many as 1% obsidian projectile points?" Answering this question is simply a matter of multiplying the probabilities of a series of sequential events.

Assume that 1% of the population of projectile points we have sampled actually are made of obsidian. The probability that the first point we select for our sample will *not* be made of obsidian is 0.99. (Since 99% are not made of obsidian, 99 times out of 100 a randomly selected point will not be made of obsidian.) There is also a probability of 99% that the second point selected for the sample will not be made of obsidian. Thus, 99% of the time we will select a nonobsidian point first. If this happens, then 99% of the time we will select a nonobsidian point second. Thus 99% of 99% of the samples of two from this population will not include obsidian points. The probability of drawing a sample of two with no obsidian points from a population with 1% obsidian points, then, is $(0.99)(0.99) = 0.980$, or 98%.

If we repeatedly select samples of two from this population, then, 98.0% of those samples will contain no obsidian points. Having found no obsidian points in a sample of two, we might continue to enlarge the sample by selecting a third random point. This third random projectile point, like any randomly selected point, will not be made of obsidian 99% of the time. Thus, in repeatedly drawing samples from this population, 98.0% of the time we will not find an obsidian point among the first two selected, and in 99% of those 98.0% of the instances, when we continue to select a third point we will still not have found one made of obsidian. Thus there is a probability of $(0.99)(0.980) = 0.970$ that a sample of three points from this population will not contain an obsidian point. We can continue in this fashion to select more and more points. At each step the probability from the previous step is multiplied once again by 0.99.

For any sample size, n, then, the probability of selecting a sample with no obsidian points from this population is 0.99^n. Thus, for a population with as many as

Table 19.1. Confidence Levels for Concluding That Absence from a Sample Indicates a Low Population Proportion

Population proportion	0.1%	0.5%	1.0%	2.0%	5.0%
n					
20	.020	.095	.182	.332	.642
25	.025	.118	.222	.397	.723
30	.030	.140	.260	.455	.785
35	.034	.161	.297	.507	.834
40	.039	.182	.331	.554	.871
45	.044	.202	.364	.597	.901
50	.049	.222	.395	.636	.923
55	.054	.241	.425	.671	.940
60	.058	.260	.453	.702	.954
70	.068	.296	.505	.757	.972
80	.077	.330	.552	.801	.983
90	.086	.363	.595	.838	.990
100	.095	.394	.634	.867	.994
110	.104	.424	.669	.892	.996
120	.113	.452	.701	.911	.998
130	.122	.479	.729	.928	.999
150	.139	.529	.779	.952	>.999
175	.161	.584	.828	.971	–
200	.181	.633	.866	.982	–
250	.221	.714	.919	.994	–
300	.259	.778	.951	.998	–
350	.295	.827	.970	.999	–
400	.330	.865	.982	>.999	–
450	.363	.895	.989	–	–
500	.394	.918	.993	–	–
600	.451	.951	.998	–	–
700	.504	.970	.999	–	–
800	.551	.982	>.999	–	–
900	.594	.989	–	–	–
1,000	.632	.993	–	–	–
1,200	.699	.998	–	–	–
1,400	.754	.999	–	–	–
1,600	.798	>.999	–	–	–
1,800	.835	–	–	–	–
2,000	.865	–	–	–	–
2,500	.918	–	–	–	–
3,000	.950	–	–	–	–
4,000	.982	–	–	–	–
5,000	.993	–	–	–	–
6,000	.998	–	–	–	–
7,000	.999	–	–	–	–

1% obsidian projectile points the probability that a sample of 16 would contain no obsidian points is $0.99^{16} = 0.851$. For a sample of 50 points, there would still be a 60.5% chance of selecting a sample with no obsidian points from a population with 1% obsidian points ($0.99^{50} = 0.605$). This probability of 60.5% amounts to an evaluation of significance. That is, it is the probability that a random sample of 50 may contain no obsidian points even though the population does have as many as 1% obsidian points. The opposite probability ($1 - 0.605 = 0.395$) is the confidence level at which we could say that the population from which our sample was selected has fewer than 1% obsidian points. Thus a sample of 50 with no obsidian points would give us only 39.5% confidence that it was selected from a population with fewer than 1% obsidian points.

Table 19.1 provides the confidence levels for given sample sizes (n) and given population proportions. The figures for the example just discussed can be found there by looking at the row for $n = 50$ and the column for a population proportion of 1%. The number in the table is 0.395, corresponding to a 39.5% level of confidence that the population from which a sample of 50 elements was drawn actually has fewer than 1% of whatever item of interest it was that failed to appear in the sample. A confidence level of 39.5% is, of course, not a very useful level of confidence at which to speak. To determine how large a sample of projectile points without any made of obsidian we would need in order to make this conclusion at the 95% confidence level, we can read farther down the column for a population proportion of 1% until we reach 0.95. In the row corresponding to a sample size of 300, the confidence level has finally reached 0.951. Thus, if we wanted a sample large enough to conclude at the 95% confidence level that the population had fewer than 1% obsidian points, we would need a sample of some 300 projectile points. (If we did find one or more obsidian projectile points in this enlarged sample, of course, we would simply turn back to the procedures discussed in Chapter 11 to estimate the proportion in the population and attach an error range at the 95% confidence level to this estimate.)

Table 19.1, then, can be used to determine the level of confidence at which we can conclude that something absent from a sample of a given size occurs in the population in a proportion of less than 5%, 2%, 1%, 0.5%, or 0.1%. It can also be used to determine how large a sample will be needed to conclude at a given confidence level that the population contains less than a certain proportion of some item of interest. As can readily be seen, if we need high confidence that the population proportion for an item absent from the sample is very low, then quite a large sample is required.

Chapter 20
Sampling and Reality

At the beginning of Chapter 7 I asserted that sampling was at the very heart of the statistical principles applied in this book. I hope the chapters that lie between there and here have made clearer just what that means. Whether the task is estimating the population mean or proportion, comparing means in several batches, comparing proportions in several batches, or investigating the relationship between two measurements, the logic of the approaches statisticians take involves thinking about the batches of numbers we are working with as samples from a larger population. It is this larger population that really interests us.

Sometimes this is literally and obviously true. If, for example, we excavate an entire rock shelter site and recover 452,516 pieces of lithic debitage, we might well select some kind of random sample of this debitage for detailed analysis with the objective of characterizing the entire population of 452,516 waste flakes. In this case we would have a sample of waste flakes that we would use to make statements about the population of all debitage at the site from which the sample was selected. The sampling design we used might well be rather complicated. For instance, we might want to be able to compare one stratum in the site to the others, so we might separately select a sample from each stratum. The techniques discussed in Chapters 9–11 would enable us to determine approximately how large each of these samples would need to be in order to accomplish our aims, and they would enable us to estimate means of measurements we might make and the proportions of different categories we might define in the several populations consisting of debitage from each stratum. We could attach error ranges to these estimates that would help us to know at what confidence level and with what precision we could discuss these means and proportions (Chapters 9–11). We could compare the means of the measurements in different strata using these estimates and error ranges or using t tests and analysis of variance (Chapters 12 and 13). We could compare the proportions of the categories in different strata using the estimates and error ranges or using chi-square (Chapter 14). We could evaluate the strength and significance of relationships between measurements with a regression analysis (Chapter 15). If we had ranks rather than true measurements, we could use a rank correlation (Chapter 16). We could combine samples from different strata to say things about the debitage from the site as a whole (Chapter 17). If there were some category of material that just did not appear in our sample, we could evaluate the confidence with which we could talk

R.D. Drennan, *Statistics for Archaeologists*, Interdisciplinary Contributions to Archaeology, DOI 10.1007/978-1-4419-0413-3_20,
© Springer Science+Business Media, LLC 2004, 2009

255

about its rarity in the population from which the sample came (Chapter 19). All these analyses would provide us with ways to say how much confidence we had that some observation of interest in the sample reflected something that occurred in the population from which the sample was selected as well. This is a very straightforward application of the principles in Chapters 9–18.

If we had a large sample of Early, Middle, and Late Archaic projectile points, we might well do all these same things with it – means and proportions with error ranges, significance tests of the differences between Early and Middle or Middle and Late, and so on. We have to stop and think for a minute, though, about just what population it is that we are talking about on the basis of our sample. Probably, the population is something like the large and very vaguely defined set of all projectile points made during the Early, Middle, and Late Archaic. Our interest is likely to be in identifying some kind of change from one period to the next in very general terms. We still clearly have a sample, and we can imagine the population we are talking about even if it is a fairly nebulous population. The sample has not been selected with truly random procedures, so the issue of sampling bias is highly relevant in this example, unlike the previous one (Chapter 7).

If we excavated a Formative village in its entirety, recovering information about 27 house structures from Early, Middle, and Late Formative, the same list of statistical tools might be put to use. If we had excavated all the houses at the site, though, it is even less clear what sense it makes to talk about these houses as a sample. What is the population they are a sample from? Aren't they the complete population? And does this mean we can't investigate the significance of a difference we might observe between, say, Early and Middle Formative? In a case like this, there are several kinds of populations we might implicitly be interested in. One is the population of all houses that existed at the site at any point in the Formative. Some have surely been destroyed by the construction of subsequent houses and other processes. Our sample is not this complete population, but in some contexts this would be the complete population of interest to us. In other contexts, we might use the sample of excavated Early Formative houses from this site as a way of talking about Early Formative houses in general. The relation between our sample and the population of interest in this context is similar to the first example concerning Archaic projectile points.

If we surveyed a whole region completely, with 100% coverage, it would become even more difficult to identify just what population we take our sample to represent. Presumably some sites would have been destroyed or made inaccessible to survey, but if conditions were so propitious for survey that we recovered data on almost all the sites, talking about our sites as a sample from a larger population has become very forced. What would it mean to talk about, say, the significance of a difference in mean site size between the Neolithic and the Bronze Age? We could certainly perform the calculations necessary to say that the mean site area in the Neolithic was 1.4 ± 0.2 ha at the 95% confidence level and in the Bronze Age it had changed to 3.6 ± 0.3 ha. (Or, instead, we could perform a t test and find out that the significance of the difference in mean site area between Neolithic and Bronze Age sites was very high.) We would thus have very high confidence that Bronze Age sites were substantially larger on average than Neolithic ones. But it is not easy to say exactly what this means in terms of samples and populations. Literally, we have concluded that it is extremely likely that our sample of Bronze Age sites came from a population of larger sites than our sample of Neolithic sites. But those two populations may really be imaginary. If our survey was so effective that unstudied sites within the region are virtually nonexistent, then the population within the region is not substantially different from our sample. It may not be any more meaningful to think of this population existing outside the region either, since what was going on in the next region may well be entirely different – perhaps there were no Neolithic sites at all.

The population we have sampled in a case like this is truly imaginary. Thinking of the things we study as samples from imaginary populations may not sound like a very good way to approach reality, but it can indeed be meaningful to talk about significance and confidence even when the things we have studied do comprise the entire population we are interested in. What an evaluation of significance like this last example tells us, in real-world terms, is that the quantity and character of the observations we have made give us a real basis to discuss a change in mean site area from Neolithic to Bronze Age. The change we have observed is very unlikely to be due to the small number and equivocal nature of our observations. In short, we have enough information to say quite confidently that something changed and to proceed to consider more fully the nature of the changes and the forces that produced them. Such an indication enables us to put to rest one of the continual worries of the archaeologist: whether we have enough information to say anything at all. Whatever other doubts we may have about our conclusions, at least we do not need to worry that we did not recover information about enough sites to tell whether or not there was a change in site size between Neolithic and Bronze Age.

Returning to think about Early, Middle, and Late Archaic projectile points, suppose that we discover that the proportion of corner-notched points in the Early Archaic is $46\% \pm 23\%$ at the 95% confidence level, and that the proportion of corner-notched points in the Middle Archaic is $34\% \pm 19\%$ (also at the 95% confidence level). We would not be much interested in talking about what caused a change in the proportion of corner-notched points, because the quantity and character of the observations we have made does not make us very confident that there *was* a change. What we might be interested in doing is visiting more museums and observing more points. With a larger sample we would be able to achieve smaller error ranges at the 95% confidence level. Eventually we should be able to see either that there was a change that we could talk about with enough confidence to make the conversation worthwhile, or, alternatively, that whatever change occurred was so small as not to be very interesting.

In either of these two cases, application of the statistical notion of confidence (or its mirror image, significance) has told us whether the quantity and character of the observations we have made are sufficient to make some conclusions. Statistical reasoning provides us with powerful tools to deal with this concern. If you find that a difference between two sets of observations you have made has high statistical significance, then at least you know that you do not have to worry about not having enough observations on which to base conclusions. To say (as some have) that such a difference has high statistical significance, but that a larger sample is needed for us to be confident about it, is to reveal lack of comprehension of what a significance test means. At the very least, the high significance means that a larger number of observations is *not* needed.

What none of this touches is the problem of sampling bias discussed in Chapter 7. Having decided to treat the observations we have as if they were a random sample enables us to go ahead and utilize the tools of statistics. We may discover that our observations, whatever they are really a sample from, are an insufficient basis to find out what we would like to find out – that is, that whatever pattern we observe has very little significance. Or we may discover that the observations we have are sufficient to tell us some statistically highly significant things about whatever population it is that they are a sample from. When we arrive at this latter point, we are put in the position once again of considering what it is we have a sample from. If the things we observed were selected in a biased manner that affects the nature of the observation we are interested in, then we must reason our way around that problem as best we can. The assistance we get from statistics in that task is limited to helping us to delineate the problem of sampling bias clearly and, in a sense, compartmentalize it by separating it analytically from the issue of sample size, which the statistical tools in this book are designed to deal with directly.

Radiocarbon dating provides a context in which archaeologists are accustomed to reasoning in this way, although we do not usually talk about it quite like this. When a sample of carbon is submitted to a dating laboratory that tells us its age is 800 ± 100 years, we say that there is about a 66% chance that the sample is between 700 and 900 years old (since by convention the error ranges expressed with radiocarbon dates are one standard error). Before we can use that result to conclude anything at all about the date of a particular stratum, there are several other issues to consider. How confident are we that the sample was uncontaminated? That it did not fall down a rodent burrow from a more recent stratum? That it was not simply a burned root of a much more recent tree? That it was not from the long-dead heartwood of a tree already ancient at the time it was deposited? In effect, all these additional questions are questions about the real nature of the population of carbon atoms of which those counted to the laboratory were a sample.

The error range tells us whether the observations are sufficient to tell us what we need to know. Suppose that what we need to know is whether or not the stratum in question is more than 400 years old. A radiocarbon age of 800 ± 100 years gives us high confidence that the sample dated is more than 400 years old. Submitting a larger chunk of carbon to the laboratory in the hope of narrowing the error range would not be at all helpful. Increasing the size of the sample would be a waste of time and money. We are quite confident that those carbon atoms were a sample from a population of carbon atoms assembled more than 400 years ago. The sample was of quite sufficient size to tell us that. The worries about how the sample was selected (in effect, about sampling bias), however, remain. Larger samples and more statistics will not help us resolve those worries. We must reason our way through those difficulties as best we can with recourse to considerations other than statistics.

The approach taken in this book to the problem of sampling bias is similar to what has become customarily accepted practice in handling radiocarbon dates. Instead of treating random sample selection as a criterion that must be met before using the tools of statistics (which, if taken literally, would stop us dead in our tracks in almost every potential application of statistics to archaeology), the issue of what our observations are really a sample from becomes a consideration in what we can conclude from our observations – if they have enough statistical significance to make it possible to conclude much from them at all. If our observations do not have enough statistical significance to make them meaningful, then we do not waste time addressing the issue of bias in the process of sample selection. What we want instead is another (larger) sample, selected in as unbiased a manner as possible.

Making conclusions or interpretations, as discussed in Chapter 7, carries us beyond the realm of statistics, although statistical tools can help put us in better position to make conclusions. The significance or confidence levels we arrive at with statistical tools have to do with the probable nature of the populations (real or imaginary) that the things we observe are samples from. Knowing how much confidence to place in our observations (or equivalently, how much significance they have) helps us evaluate just how much support they provide for the conclusions we are interested in making. The confidence probabilities, however, do not apply directly to the conclusions.

If we think that family size increased from one period to the next and that such a change might result in larger storage jars in the second period, we can use statistical tools to determine how much confidence we should have in our observation that storage jar volume increased. If we find that we are 99% confident that storage jar volume increased, that does not mean that we are 99% confident that family size increased. There are other possible reasons for the increase in storage jar volume. The confidence probabilities pertain to our observations, not to what we think our observations mean. The high confidence we have in our observation of storage jar volume increase, of course, provides evidence in favor of our conclusion. (If our calculations gave us less confidence that storage jar volume increased, then our observations would provide less support for our conclusion that family size increased.) Observations at high confidence levels of other kinds of evidence consistent with increases in family size would add more support to our conclusion, while observations at high confidence levels of other kinds of evidence inconsistent with increases in family size would make us doubt our conclusion. The weighing together of these multiple, quite possibly contradictory, observations of different lines of evidence is essential to the process of evaluating conclusions, and it is not primarily a statistical task.

Statistical tools are more useful at a lower level in helping us to evaluate each individual line of evidence and to assess just how much (or how little) support each set of observations contributes toward sustaining our ultimate conclusions. It is because we use statistics most often in a context like this rather than in a context where we must make yes or no decisions based on observation of a single variable in a single sample that a scalar rather than null hypothesis testing approach to framing significance tests is particularly suitable in archaeology. Using samples simply to make estimates with error ranges about the populations they were selected from and comparing these estimates with bullet graphs is, in many instances even clearer and more direct.

In particular, the statistical tools most discussed in this book provide a powerful approach to the archaeologist's perennial worry about whether there are enough observations to conclude anything much. Are the 253 sherds collected from this site enough to enable us to talk very confidently about proportions of different types? How confidently can we discuss the size of Formative houses on the basis of the five house floors we have excavated? Do we have much confidence that the number of temple mounds depends on site area, given that our observation of this relationship is based on only 16 sites? Do we need to analyze all the unifacial flake tools recovered from this site, or could we learn more by studying a sample intensively? If a sample, how large would it need to be to tell us what we need to know? These are all examples of questions that have loomed large in archaeologists' sleepless nights for decades. They have been answered too often in archaeological research in purely arbitrary and subjective ways. Answering such questions on the basis of subjective impressions or "gut feelings," is unsupportable. It is precisely these kinds of questions that the statistical tools explored thus far in this book can help us whittle down to size.

Part V
Multivariate Analysis

Chapter 21
Multivariate Approaches and Variables

What we did in Part III (Relationships between Two Variables) is often labeled *bivariate analysis*. The chapters that form this last part of the book extend the investigation of relationships between two variables into the *multivariate* arena. There are many different approaches to multivariate analysis. Those discussed here represent only a very small selection, one that emphasizes techniques particularly suitable for exploration that seeks patterns in multivariate datasets (as contrasted to evaluating the strength and significance of particular patterns that are hypothesized in advance). The perspective of this final section of the book, then, strongly recalls the exploration of batches in Part I. The exploration here, though, is not of single batches of numbers, or variables. It goes on beyond relationships between two variables to approach directly the more complicated situation in which we have a larger number of variables for each case. We could, of course, approach a large number of variables piecemeal by looking at their relationships by pairs, taking them two at a time, and evaluating the strength and significance of the relationships between them with the tools discussed in Part III. This would soon get out of hand without some way to truly combine the results of all those pairwise evaluations of relationships.

In Chapter 15 we actually discussed one way to accomplish just this, using the residuals from one regression analysis as the dependent variable in another regression analysis. This is, in fact, the avenue toward one form of multivariate analysis. Multiple regression does just this in an integrated way as a single analysis, and most statpacks will perform multiple regression. Since its basic principles are precisely those described in Chapter 15 for integrating several bivariate regression analyses into a single set of results, multiple regression will not be further discussed here. Multiple regression has, however, been used in archaeology, especially in situations where the aim is to use a number of variables together to predict the value of a single important dependent variable. Models for predicting site locations, for example, have often made heavy use of multiple regression. Multiple regression differs from the multivariate techniques we will discuss in another way as well. Like bivariate regression, multiple regression evaluates the strength and significance of a particular

R.D. Drennan, *Statistics for Archaeologists*, Interdisciplinary Contributions
to Archaeology, DOI 10.1007/978-1-4419-0413-3_21,
© Springer Science+Business Media, LLC 2004, 2009

kind of relationship between variables – a relationship expressed in an equation for predicting the value of a dependent variable, based on the values of a series of independent variables. It is thus less an exploration for patterning than an evaluation of how well a particular hypothetical way of expressing relationships works on a given dataset.

The context we will focus on here is a more exploratory one, in which there is not a single major dependent variable whose value we seek to predict, nor a specific hypothesized model of relationships between variables that we seek to evaluate. Instead, we will consider a set of things with a number of varying characteristics encapsulated in a set of measurements and/or categories. Our aim with the tools discussed in these final chapters is to find patterns of relationships in such a dataset – relationships that can be expressed in several different ways.

A concrete example may make some of this vague talk a bit clearer. One of the central reasons archaeologists got interested in multivariate analysis in the first place was a desire to make the definition of artifact typologies more rigorous and replicable. The creation of, say, a ceramic typology can easily be considered a multivariate analysis problem. The traditional way of creating a ceramic typology is, of course, to put a lot of sherds on a very large table and push them around into piles that bring together sherds that are similar to each other and separate those that are different. Once the pushing around is done, the characteristics of the sherds in each pile are written down in what becomes a type description. This has often seemed a capricious and arbitrary practice, and artifact classification in the archaeological laboratory is legendary for disagreements about just how to make the piles or just which pile to put a particular sherd in.

There was a moment when it seemed to some archaeologists that multivariate analysis offered the solution to these difficulties. The large number of sherds could become cases in a dataset; and each of their potentially interesting and useful characteristics, a variable. Some of the variables are categories (such as surface finish, which might be either rough, smoothed, or burnished); others could be measurements (such as rim thickness in mm). *Voilà*, a multivariate dataset! The right statistical analysis, then, could ferret out the patterns in the multivariate dataset and banish forever all this quibbling over artifact types. For several reasons, it has not worked out quite this way, and it is not in the effort to make artifact typologies more rigorous and replicable that multivariate analysis now has its most promising uses in archaeology. The basic idea, though, of using multivariate analysis to improve artifact typologies helps make clear just what a multivariate dataset is like, and just what it might mean to explore for patterning in it.

A SAMPLE DATASET

We will use the same multivariate dataset as a continuing example through all of the following chapters. Like the other datasets used in this book, it is made up so as to help us focus more clearly on central principles. The patterns in it are fairly simple and straightforward, but they are not perfect. The dataset contains enough random noise and general messiness to provide a more or less realistic opportunity to see how different multivariate techniques present patterns in real datasets.

The cases in the dataset we will use are 20 household units from the fictitious Formative Mesoamerican site of Ixcaquixtla. The excavations at Ixcaquixtla revealed the residential structures corresponding to these 20 households and recovered substantial samples of artifacts from middens associated with each one. A remarkable number of burials were also encountered in the area around each house. The site was occupied for only a short period of time, making it possible to treat these household units as contemporaneous, and they apparently comprise the majority of the households that existed at this small village site during its period of occupation.

The data are presented in Table 21.1. The household units have been assigned numbers to facilitate keeping track of which is which, and ten variables encapsulate some of the characteristics to be observed in the house structures, the burials, and the artifact and ecofact assemblages from the middens. The first variable, *Bowls %* *of Sherds*, is the proportion of the sherds recovered from the midden deposits that can be identified as serving bowls. Most of the rest of the sherds are from storage and cooking vessels, so a high proportion indicates a large amount of finer food-serving vessels in an assemblage. It is not uncommon to relate a high proportion

Table 21.1. A Multivariate Dataset on Households at Ixcaquixtla

Household Unit	Bowls % of Sherds	Decoration % of Sherds	Energy Invested in Burials	Mace Heads	Fauna/ Sherd Ratio	Platform	Shell/ Sherd Ratio	Wasters % of Sherds	Debitage % of Lithics	Obsidian % of Lithics
1	0.25	0.03	2	0	0.32	0	0.000	0.000	0.79	0.00
2	0.37	0.07	3	0	0.55	0	0.000	0.000	0.35	0.00
3	0.15	0.01	1	1	0.10	1	0.008	0.000	0.32	0.00
4	0.19	0.01	2	0	0.20	0	0.000	0.000	0.26	0.00
5	0.35	0.04	3	0	0.57	0	0.000	0.000	0.69	0.00
6	0.21	0.01	1	0	0.13	1	0.000	0.000	0.31	0.12
7	0.24	0.01	1	0	0.19	0	0.000	0.000	0.86	0.00
8	0.20	0.05	2	0	0.28	0	0.000	0.016	0.19	0.00
9	0.49	0.09	3	0	0.48	0	0.000	0.000	0.28	0.00
10	0.23	0.02	2	0	0.24	0	0.000	0.000	0.29	0.00
11	0.26	0.02	2	1	0.21	1	0.000	0.021	0.31	0.13
12	0.19	0.00	1	0	0.15	0	0.000	0.000	0.46	0.00
13	0.31	0.04	2	0	0.37	1	0.000	0.025	0.26	0.10
14	0.45	0.05	3	1	0.60	1	0.009	0.000	0.65	0.00
15	0.48	0.03	3	0	0.43	0	0.000	0.000	0.43	0.00
16	0.09	0.00	1	1	0.15	0	0.005	0.000	0.29	0.00
17	0.11	0.02	1	0	0.09	0	0.000	0.014	0.28	0.00
18	0.29	0.02	2	1	0.25	1	0.007	0.000	0.87	0.00
19	0.28	0.03	2	1	0.40	0	0.000	0.000	0.31	0.00
20	0.19	0.03	1	0	0.05	0	0.000	0.000	0.95	0.00

of such ceramics to more elegant or elaborate food serving, as may occur in elite households or feasting. The second variable, *Decoration % of Sherds*, is the proportion of the sherds recovered from the midden deposits that have elaborate painted designs. These painted vessels clearly required considerably more labor to produce than undecorated ones, so their possession might be related to wealth or prestige.

Burials were located around the houses in which the deceased had apparently lived. Sometimes burials were placed in plain excavated grave pits and sometimes in more elaborate stone slab tombs, and the quality and quantity of offerings included with them varied substantially. *Energy Invested in Burials* is a rough categorization of the labor required on average for the burials associated with each household unit. This could not be calculated at all precisely, but was rated in three categories (1 = low, 2 = medium, and 3 = high). Investment of large amounts of labor, on average, in the burials of the deceased members of a particular household is often taken to indicate something special about that household – whether economic wealth, social prestige, political power, supernatural authority, ritual position, or something else. *Mace heads* are elaborately carved stones that seem likely to have been more ceremonial than practical. They were included in burials occasionally, but were very rare, so only their presence (1) or absence (0) is recorded.

The *Fauna/Sherd Ratio* is the number of animal bones recovered from the midden deposits associated with each household unit divided by the total number of sherds recovered from those same deposits. In some contexts, meat is taken to be a preferred food, and households able to consume more meat, as indicated by concentrations of faunal remains, are regarded as special in some way. Some house structures were built atop plastered *Platforms* about 1 m high. A value of 1 for this variable indicates that the house was on a platform; a value of 0, that it was not. Again, house structures raised conspicuously on platforms are likely to be taken to indicate something special about that household – high rank in a social, economic, or political hierarchy or perhaps differentiation based on some less hierarchical principle.

The *Shell/Sherd Ratio* is the total number of marine shell fragments recovered from the midden deposits associated with each household divided by the total number of sherds recovered from those same deposits. Marine shell had to be imported from distant coastal regions, so it was rare and apparently of ornamental use. Most of the fragments found in the middens appear to be not finished ornaments, but debris from making such ornaments. An archaeological interpretation is thus likely to relate them to craft activity. Ceramic production is indicated by the presence of kiln wasters – fired ceramic fragments that were so badly damaged during the manufacturing process that they would never have been useful and would just have been discarded straight away. *Wasters % of Sherds* is the total number of such kiln wasters recovered from the midden deposits associated with each household unit divided by the total number of sherds recovered from those same deposits. Waste flakes from the manufacture of flaked stone tools were recovered in fairly large quantity from fine screening of the midden deposits. The number of such waste flakes recovered from the midden deposits associated with each household divided by the total number of flaked stone artifacts recovered from those same deposits is recorded as *Debitage % of Lithics*. Finally obsidian was an especially high-quality raw material for flaked stone tools. It was imported from one of a restricted number of obsidian sources, all of which lie at long distances from Ixcaquixtla. *Obsidian % of Lithics* records the total number of fragments of obsidian recovered from the midden deposits associated with each household unit divided by the total number of flaked stone artifacts recovered from those same deposits.

This multivariate dataset, then, includes a variety of kinds of information about household units at Ixcaquixtla. The variables are kinds of evidence that have often been related to subjects such as status, wealth, leadership, feasting, ritual, and craft production. We might well be interested in exploring such data for patterns, whether we had any particular hypothesis in mind to evaluate or not. This is precisely the exploration we will undertake in the next few chapters by several different approaches to multivariate analysis.

KINDS OF VARIABLES, MISSING DATA, AND STATPACKS

There are several kinds of variables in Table 21.1, and since they may receive different treatments in some multivariate approaches, it is necessary to pay attention to a few more details about kinds of variables than we have distinguished up to now. Most of the variables in Table 21.1 are measurements (the proportions and ratios). Two of the variables (Mace Heads and Platforms) are categories. The categories are *mutually exclusive and exhaustive*, as all sets of categories must be. That is, each case must fit into one and only one of the categories for each variable. This has been true of all the category variables we have dealt with in earlier chapters; it seems so self-evident in that context that it was not even necessary to mention it. The urge to define variables that do not have this characteristic does sometimes arise in multivariate analysis, however, and it must be resisted.

For each household, clearly Mace Heads are either present or absent among the burials; they cannot be both, and they must be one or the other. The same is true of the variable Platform. Presence/absence variables like these two sometimes receive special treatment in multivariate analysis, distinct from other sorts of category variables. A dataset like this might also have a variable, say, Wall Construction, whose three categories were *wattle-and-daub*, *wood-plank*, and *mud-brick*. These categories would also need to be mutually exclusive and exhaustive. If both wood plank and mud brick wall construction were found in a single house structure, a fourth category (*wood-plank-with-mud-brick*) would need to be added, so that each household unit could be placed in one and only one category. This category variable would not, however, be a presence/absence variable, and in some approaches would not be treated in the same way as a presence/absence variable.

Finally, there is one variable in Table 21.1 that represents ranks (Energy Invested in Burials). In Chapter 16 we treated ranks as a sort inferior measurement variable – one where higher values do mean "more" but where a value twice as high as another cannot be taken to mean "twice as much," as is the case with real measurements. Here Energy Invested in Burials takes the form of a variable with three categories, and is, in this respect, much like the other category variables. It differs from the other category variables in that the three ranks definitely come in a prescribed order. We recognize that to say low, high, medium is to put the categories out of their correct order. When we assign numbers to them, then, we assign 1 to low, 2 to medium, and 3 to high so that mathematical manipulations of those numbers can recognize that in a very meaningful sense medium falls between low and high.

The data in Table 21.1 are organized as they might need to be for a statpack. The measurements, of course, are represented by their numeric values, as always. The categories are also now represented by assigned numeric values, something we have not needed to do previously with categories. It is necessary in a multivariate dataset because the values of the variables must be manipulated mathematically. We cannot assign, say, "P" to present and "A" to absent for Platform, because we cannot add, subtract, divide, and multiply with "P" and "A." In principle, we could assign a value of 0 to present and a value of 1 to absent, but it is customary to do it the other way around. It is easier to use most software that recognizes differences between presence/absence variables and other kinds of variables if 0 is the value of absent and 1 is the value of present.

Values of 1, 2, and 3 have been assigned to Energy Invested in Burials, just as we have done previously with ranks. In principle, we could have assigned 3 to low, 2 to medium, and 1 to high, but it is much less confusing to assign low numeric values to low amounts and high numeric values to high amounts. If the dataset had a category variable like Wall Construction (wattle-and-daub, wood-plank, mud-brick, or wood-plank-with-mud-brick), we would assign numbers to each of these categories as well. Ordinarily we would not use 0 as one of the values for this variable, since none of the categories really means absence. Instead we might use 1 for wattle-and-daub, 2 for wood-plank, 3 for mud-brick, and 4 for wood-plank-with-mud-brick. We could easily mix the number values around, though, for this variable. Any of the four categories might be assigned a value of 1 since the number values do not represent any sense of ranking of the four categories. It must be remembered that the numbers assigned to the categories for Energy Invested in Burials convey information about ranks, while those assigned to the categories for Wall Construction would not. This distinction can matter in multivariate analysis.

The notion of *missing data* also plays an especially important role in multivariate analysis. We have not needed to be concerned about it previously because missing data usually takes care of itself when dealing with one variable or two. If a scraper is broken, and we cannot measure its length, then automatically no measurement for it appears in the batch of numbers we are exploring and calculating indexes for. That scraper just disappears from the sample when we look at length measurements. It might well reappear when we look at the batch consisting of categories of raw material. The fact that it is broken would not prevent identifying the raw material of which it was made. If we investigated the relationship between scraper length and raw material, the broken scraper would disappear again. It would disappear because we would have no measurement for its length and could not include that case in the sample for analysis. We could have said much earlier on that such a case was excluded from the analysis by reason of missing data, the missing data being the missing length measurement.

In multivariate datasets, it is quite common to be able to measure or classify all the cases for most of, but not all, the variables. For example, if there were a household unit in the dataset in Table 21.1 that had no burials associated with it at all, we would not be able either to categorize the average Energy Invested in Burials or say whether Mace Heads were present or absent in burials. This household unit would need to be assigned a special value for those two variables – a value that reflected this special condition. All statpacks have an established procedure for dealing with missing data. A particular value, not otherwise used, is likely to be established as a *missing data code*. It may be a period all by itself (".") or some other character not ordinarily used in recording data. In multivariate analysis it becomes of great importance to use such missing data codes effectively, and to distinguish clearly between "missing data" and "absent." It is sometimes necessary to choose between different ways in which cases with missing data are dealt with by your statpack.

The concept of missing data applies to both measurements and categories. If, for example, the sherds recovered from the midden deposits associated with one of the Ixcaquixtla households were all so disastrously eroded that it was impossible to tell whether they had ever been decorated or not, the appropriate value to use for Decoration % of Sherds would not be 0.00 but "missing data." If some of the sherds were so badly eroded that decoration, if present, would not be discernible, then those sherds would be excluded from the counts upon which Decoration % of Sherds is based. Badly eroded sherds would be excluded both from the numerator and the denominator of this fraction, so that the proportion would be calculated by dividing the number of sherds with decoration by the number of sherds well enough preserved that decoration would be noticed if it had been present.

Chapter 22
Similarities between Cases

Several approaches to multivariate analysis begin by assessing the similarities of each case in a dataset to each other case in the dataset, basing the measure of similarity on the values of the set of variables that have been recorded for each case. Such measures of similarity are called *similarity coefficients*. The notion of similarity between cases in this instance is exactly what common sense implies. Two cases are quite similar if they have similar values for each of the variables measured and less similar if they have rather different values for each of the variables. Actual objects with physical measurements provide a clear illustration, for example the projectile point measurements in Table 22.1. Looking first at the length measurements for the four projectile points, we would easily recognize that Points 1, 2, and 3 are quite similar with regard to length, while Point 4 is rather different. Looking at thickness, however, we would equally quickly say that it is Point 2 that stands out as different from the group. Yet a different pattern emerges with regard to weight: here Points 1 and 4 are quite similar (identical in fact), and Points 2 and 3 are identical to each other but different from 1 and 4.

In short, the differences and similarities between the projectile points are easily observed with regard to single variables. When we wish to consider all variables at once, however, the situation rapidly becomes much more complex. If asked which two projectile points were most alike, considering all the variables, we would have no immediately obvious answer. This is the situation that similarity coefficients are designed to deal with. Similarity coefficients are indexes of how similar two cases are, considering simultaneously all the variables for which they have been measured. The larger the similarity coefficient, the more similar the cases are.

Similarity coefficients sometimes come in the form of *dissimilarity coefficients*, or numbers that are larger when two cases are less similar (that is, more dissimilar).

R.D. Drennan, *Statistics for Archaeologists*, Interdisciplinary Contributions
to Archaeology, DOI 10.1007/978-1-4419-0413-3_22,
© Springer Science+Business Media, LLC 2004, 2009

Table 22.1. Measurements for Four Projectile Points

Point No.	Length (cm)	Width (cm)	Thickness (cm)	Weight (g)
1	4.3	1.2	0.35	75
2	4.5	1.4	0.55	80
3	4.4	1.1	0.37	80
4	2.3	0.9	0.30	75

Some dissimilarity coefficients are referred to as *distances*, indicating metaphorically how far apart two cases are on a scale of similarity. The distinction between similarity coefficients and dissimilarity coefficients (or distances) is a trivial one in principle, but a crucial one in practice. It is a simple proposition to convert a similarity coefficient into a dissimilarity coefficient, or vice versa, by simply subtracting either one from the maximum value it takes on. It is also easy to accidentally use dissimilarity coefficients as if they were similarity coefficients by specifying the wrong option in your statpack. The result would be to turn all the relationships in a dataset on their heads and produce nonsense in a multivariate analysis.

There are a number of different similarity coefficients that are suitable for different kinds of variables. Some were designed with measurements in mind; some, with categories; and some, specifically with presence/absence variables.

EUCLIDEAN DISTANCE

One of the most frequently used and versatile coefficients for measuring similarity between cases is *Euclidean distance*, based on the familiar Pythagorean Theorem of grade-school geometry. If we consider for a moment just the two variables, length and width in Table 22.1, it is easy to create a scatter plot as in Fig. 22.1 representing the length on the x axis and the width on the y axis. The distances between the points in this graph are a good commonsense representation of the dissimilarities between the four projectile points. Points 1, 2, and 3 are relatively close together, while Point 4 is set clearly apart from them. When we look at the lengths and widths in Table 22.1, we see that this makes good sense; Points 1, 2, and 3 have relatively similar lengths and widths, while Point 4 is rather different from this group, especially in regard to length. We could simply measure or calculate the distance shown in Fig. 22.1 and use it as a measure of how dissimilar the projectile points are with regard to the two variables length and width.

The distance between Points 1 and 4 in Fig. 22.1 can be calculated using the Pythagorean Theorem. The length of the horizontal leg of the right triangle is the difference in length between the two points (or 4.3 cm − 2.3 cm = 2.0 cm). The length of the vertical leg of the right triangle is the difference in width between the two points (or 1.2 cm − 0.9 cm = 0.3 cm). The straight-line distance between Point 1 and Point 4 is the length of the hypotenuse, which is the square root of the sum of the

Figure 22.1. Measurement of Euclidean distance between projectile points in two dimensions.

squares of the two legs, or

$$\sqrt{2.0^2 + 0.3^2} = \sqrt{4.0 + .09} = 2.02$$

We could calculate distances in this way between each pair of points, finding that Points 1 and 2 are relatively close together, as are Points 1 and 3 and Points 2 and 3. Points 1 and 4, 2 and 4, and 3 and 4 are substantially farther apart. These are the Euclidean distances between each pair of points in simple Euclidean geometry on this two-dimensional plane defined by the two variables, length and width.

Precisely the same logic for any number of variables, and the same calculation of distance can be made. It is easy to visualize adding a third variable and a corresponding z axis to the graph. It would become a three-dimensional graph with its z axis perpendicular to the page. In algebraic terms, the distance between any pair of points would simply be the square root of the sum of the squares of the differences between the values for the two cases on each variable. This algebra is extendable to any number of dimensions, and the formula for Euclidean distance becomes

$$D_{1,2} = \sqrt{\sum \left(X_{j,1} - X_{j,2} \right)^2}$$

where $D_{1,2}$ = the Euclidean distance between cases 1 and 2, $X_{j,1}$ = the value of the jth variable for case 1, and $X_{j,2}$ = the value of the jth variable for case 2.

The squared differences are summed for all j variables, resulting in a single distance for each pair of cases, considering all the variables at once. The resulting table of distances appears in Table 22.2. Such a matrix is often referred to as a *square symmetrical matrix*. It will always be square since there is a row for each case and a column for each case, and thus the number of rows is always the same as the number of columns. It will always be symmetrical since the distance or dissimilarity between Case 1 and Case 2 is always the same as the distance or dissimilarity between Case 2 and Case 1 (for Euclidean distances at least). The table is, in fact, just like a table

**Table 22.2. Euclidean Distances between Projectile Points
from Table 22.1**

	1	2	3	4
1	0.0000			
2	5.0120	0.0000		
3	5.0020	0.3639	0.0000	
4	2.0230	5.4911	5.4272	0.0000

of distances between cities in the margin of a highway map. Along the diagonal of such a table running from upper left to lower right, the values in all the cells are zeros, since they represent the Euclidean distances between each case and itself, which is always 0. The values in all the cells above and to the right of this diagonal simply mirror the values below and to the left of this diagonal, since the two halves of the table represent the same distances measured in opposite directions. Because of this, tables of similarities or dissimilarities are often printed or stored as triangular half-tables to save space. This is the practice followed in Table 22.2. According to this table of distances or dissimilarities based on all four measurements, Points 2 and 3 are the most alike because they show the smallest distance or dissimilarity (0.3639), and Points 2 and 4 are the least alike because they show the largest distance or dissimilarity (5.4911).

EUCLIDEAN DISTANCE WITH STANDARDIZED VARIABLES

Upon reflection, there are reasons to be dissatisfied with the indexes of dissimilarity in Table 22.2. Points 2 and 3, with by far the shortest Euclidean distance, certainly are similar in regard to weight and length, but their differences in regard to width and thickness do not seem adequately to have been taken into account. The Euclidean distance between Points 1 and 2 is 5.0120, far greater than the distance between Points 2 and 3, but if we look across all four variables, it does not seem reasonable that we should consider 1 and 2 so much more different from each other than 2 and 3 are.

In Table 22.1, the length difference between Point 4 and the others is the most notable difference of all. It only amounts to about 2 cm in length, but the longer points are nearly twice as long as the shorter one, and this calls our attention to the difference, as well it should. We would likely consider the difference in weight between Points 3 and 4 to be minor when compared with the difference in length between Point 4 and the others. These observations are implicitly based on the notion of unusualness that we have been working with since Chapter 4. Considering the lengths of the four projectile points, Point 4 is quite unusual. Its length of 2.3 cm is 1.5 standard deviations below the mean length of 3.88 cm. Point 4 differs from the other three points in regard to length by anywhere from 1.9 to 2.1 standard deviations. The weight of 80 g for Point 3, however, while different from Point 4, is not so unusual as is the length of Point 4. This weight of 80 g is only 0.87 standard deviations above the mean weight of 77.5 g. The points weighing 80 g differ from the points weighing 75 g by 1.7 standard deviations. Here a difference of 5 g in weight is less unusual (1.7 standard deviations), and thus matters less to us, than a difference of only 2 cm in length (1.9–2.1 standard deviations). That a difference of 5 matters less than a difference of 2 here does not surprise us; after all length measured in centimeters and weight measured in grams are on inherently different scales that cannot be compared to each other meaningfully in this way. Our calculation of Euclidean distance, however, has treated the difference of 5 g to be much larger than the difference of 2 cm because 5 is much larger than 2. In calculating Euclidean distance, we implicitly treated each of these scales as if they were fully comparable.

The fact that one is a measurement of length and the other a measurement of weight, however, is only part of the story. A much more fundamental aspect of incomparability applies even to measurements on the same kind of scale in the same units. Projectile Points 1 and 2 differ from each other by 0.2 cm in length and by exactly the same amount in width. The difference in width, however, is a difference that matters much more since it is a difference of 1.0 standard deviations of width. The difference in length (of exactly the same 0.2 cm) is a difference of only 0.2 standard deviations of length.

Both these aspects of incomparability of scales can strongly affect the calculation of Euclidean distance. It is usually a good idea to base the calculation of Euclidean distance on measurements expressed in terms of their unusualness in their own respective batches rather than on their original units of measurement. We can use the customary way of re-expressing a batch of measurements on a scale of unusualness by removing the level and spread. In calculations of Euclidean distance this is usually done by *standardizing* with the mean and standard deviation. That is, for each variable, the mean of the batch for that variable is subtracted from each number in the batch, and the remainder is divided by the standard deviation. The standardized variables from Table 22.1 are given in Table 22.3. The length of Point 1, for example, is 0.404 standard deviations longer than the mean projectile point length, and Point 4 is 1.495 standard deviations shorter than the mean projectile point length.

Table 22.3. Standardized Measurements for Four Projectile Points

Point No.	Length	Width	Thickness	Weight
1	0.404	0.240	−0.390	−0.866
2	0.593	1.201	1.444	0.866
3	0.498	−0.240	−0.206	0.866
4	−1.495	−1.201	−0.848	−0.866

Table 22.4. Euclidean Distances between Projectile Points from
Table 22.1 Based on Standardized Variables in Table 22.3

	1	2	3	4
1	0.0000			
2	2.7061	0.0000		
3	1.8093	2.1933	0.0000	
4	2.4276	5.2882	2.8829	0.0000

Euclidean distances are then calculated on these numbers in exactly the same way we first calculated them on the unmodified measurements. The standardization has changed the coordinate system of the imaginary multidimensional space within which we calculate the distances. Instead of an axis measured in centimeters that corresponds to projectile point length, we have an axis whose units are in standard deviations of length above and below the mean length. The same thing happens to each of the axes (variables). The Euclidean distances between each pair of projectile points, based on standardized measurements, are given in Table 22.4.

Points 2 and 3, which were separated by such a short distance before the measurements were standardized, are now separated by a much larger distance. The important differences between 2 and 3 in regard to width and thickness which counted for so little before, now count much more heavily. They counted for so little before because the raw numbers for width and thickness are so much smaller across the board than the raw numbers for length and weight. Previously the difference of 5 g in weight between Points 1 and 4 on the one hand and Points 2 and 3 on the other hand made for very large Euclidean distances between the pairs 1/2, 1/3, 2/4, and 3/4. Standardization has placed these large raw differences in weight on a scale more appropriately comparable with the much smaller raw differences in width and thickness. In the vast majority of multivariate analyses, there is much to be gained by standardizing measurements before calculating Euclidean distances, and there is seldom anything to be lost by doing it.

WHEN TO USE EUCLIDEAN DISTANCE

Euclidean distance is a measure of dissimilarity that can be used with most kinds of variables. It is most commonly used when the variables are true measurements, and it is in such a case that the calculation of Euclidean distances makes most sense (especially if the variables are standardized). It makes reasonably good sense if the variables are ranks as well, even though it will treat a rank of 4 as twice as much as a rank of 2. Even presence/absence variables (or other kinds of two-category variables) are treated meaningfully in the calculation of Euclidean distance.

The one kind of variable that poses a serious problem for the calculation of Euclidean distance is a variable with more than two unranked categories. The Wall Construction variable imagined for the Ixcaquixtla household dataset in Chapter 21 is such a variable. Its four categories of different kinds of wall construction are just all different from each other – no pair of them any more different from each other than any other pair. If the values 1–4 are assigned to the four categories, however, the calculation of Euclidean distance will inevitably treat categories 1 and 4 as more different from each other than 1 and 2. This seems undesirable enough that Euclidean distances should not be used with such variables. One solution worth considering is to reorganize the dataset, so that each of the categories is a separate presence/absence variable. There would then be three new presence/absence variables: wattle-and-daub walls, wood-plank walls, and mud-brick walls, each coded independently as present or absent for each household. Since these have become separate variables rather than mutually exclusive and exhaustive categories of a single variable, there is no need for combinations like wood-plank-and-mud-brick walls. Such a case would simply be coded present for both kinds of wall.

As noted earlier, standardization is very often a good idea in the calculation of Euclidean distances, even though calculating a mean and standard deviation for a presence/absence variable and using them to standardize it makes little sense in and of itself. Standardization does tend to equalize the impact of the different variables, and in most cases this is desirable.

PRESENCE/ABSENCE VARIABLES: SIMPLE MATCHING AND JACCARD'S COEFFICIENTS

Several special similarity coefficients have been suggested for use when all the variables consist only of two categories: present and absent. In such a situation, all the possible results of comparing two cases for a single variable are summarized in the crosstabulation of Table 22.5. Cell a in the table represents the result if the variable is present for Case 1 and present for Case 2 (sometimes called *present-present matches*). Cell b represents absent for Case 1 and present for Case 2 (a *mismatch* between the two cases). Cell c represents present for Case 1 and absent for Case 2 (another *mismatch*). And cell d represents absent for both Case 1 and Case 2 (an *absent-absent match*). A tabulation in the form of Table 22.5 can be made for all variables and for each pair of cases, such that cell a becomes the total number of present–present matches for the two cases under consideration across all variables, and so on.

The simplest coefficient based on such a tabulation is, not surprisingly, called the *Simple Matching Coefficient*. It is the total number of matches divided by the total number of variables, or

$$\frac{a+d}{a+b+c+d}$$

For example, for the data on sherds in Table 22.6, the Simple Matching Coefficient for Sherds 1 and 2 is three matches divided by the total of six variables, or 0.5000. For Sherds 1 and 3, there are also three matches divided by six variables, or 0.5000. The two most similar sherds are 6 and 7: six matches divided by six

**Table 22.5. The Four Possible Results of Comparing
Two Cases for a Presence/Absence Variable**

		Case 1	
		Present	Absent
	Present	a	b
Case 2			
	Absent	c	d

Table 22.6. Some Presence/Absence Variables Coded for a Set of Seven Sherds

Sherd No.	Slip	Red Paint	Incising	Punctations	Quartz Temper	Mica Temper
1	Present	Absent	Absent	Absent	Absent	Absent
2	Absent	Present	Present	Absent	Absent	Absent
3	Absent	Present	Absent	Absent	Absent	Present
4	Present	Absent	Present	Absent	Absent	Present
5	Present	Present	Absent	Present	Absent	Absent
6	Present	Absent	Present	Absent	Present	Absent
7	Present	Absent	Present	Absent	Present	Absent

variables, or 1.0000. Since these two sherds are identical, we can notice that the largest value the Simple Matching Coefficient ever has is 1. Its lowest possible value is 0, the result of 0 matches divided by any number of variables. Thus the Simple Matching Coefficient ranges from 0, for two maximally dissimilar cases, to 1, for two identical cases. This property, ranging from 0 to 1, is a useful one for a coefficient to have, and this is an advantage of the Simple Matching Coefficient over Euclidean distance, which ranges from 0 for no distance or dissimilarity to an indeterminately large number for a pair of cases that are very different.

Sometimes, when presence/absence variables represent categories that rarely occur (as, for example, incising, punctations, and quartz and mica temper in Table 22.6), a present–present match is considered more meaningful than an absent–absent match. That is, the fact that Sherds 6 and 7 both have quartz temper is probably a much more meaningful match than the fact that both of them do not have punctations. Most sherds, after all, do not have punctations or quartz temper, so it is not so remarkable to find two sherds that lacked both. In many regards, we tend to remark on the co-occurrence of rare characteristics more than the co-occurrence of common characteristics. It is more striking if two people meet and discover they have the same birthday than if two people meet and discover they are both right handed. *Jaccard's Coefficient* was designed with this observation in mind. It is the number of present–present matches divided by the number of present–present matches plus the number of mismatches, or

$$\frac{a}{a+b+c}$$

Jaccard's Coefficient thus completely ignores absent–absent matches as uninterest-ing, much as you might ignore as uninteresting the coincidence of meeting someone who did not have your same birthday. Jaccard's Coefficient is a sensible choice where presence/absence variables deal with rarely occurring categories.

Like the Simple Matching Coefficient, Jaccard's Coefficient ranges from 0 to 1. Both are similarities (as opposed to dissimilarities like Euclidean distances) since large values mean more similar cases and smaller values mean less similar cases. Both are typically expressed as square symmetrical matrices, as with Euclidean dis-tances, and they are often printed as lower left half matrices, including only the nonredundant numbers in one triangular half of the matrix. Sometimes the upper right half is printed instead of the lower left, but this is less common. Sometimes the diagonal appears (in the case of these two similarity coefficients, all the num-bers along the diagonal will be ones); sometimes it does not. The Simple Matching Coefficient matrix and the Jaccard's Coefficient matrix for the sherds in Table 22.6 are given in Tables 22.7 and 22.8. Comparing the two tables shows the different assessments these two coefficients provide of relationships between cases. While many pairs of cases that have high similarity scores in one table also have high sim-ilarity scores in the other, similarity relationships do change as well with the change in coefficient. In Table 22.7, Sherds 1 and 2, for example, are rated as having the same degree of similarity to each other as Sherds 3 and 4 do. In Table 22.8, Sherds 1 and 2 are rated as completely dissimilar, but Sherds 3 and 4 are not.

Table 22.7. Simple Matching Coefficient of Similarity between the Sherds in Table 22.6

	1	2	3	4	5	6	7
1	1.0000						
2	0.5000	1.0000					
3	0.5000	0.6667	1.0000				
4	0.6667	0.5000	0.5000	1.0000			
5	0.6667	0.5000	0.5000	0.3333	1.0000		
6	0.6667	0.5000	0.1667	0.6667	0.3333	1.0000	
7	0.6667	0.5000	0.1667	0.6667	0.3333	1.0000	1.0000

Table 22.8. Jaccard's Coefficient of Similarity between the Sherds in Table 22.6

	1	2	3	4	5	6	7
1	1.0000						
2	0.0000	1.0000					
3	0.0000	0.3333	1.0000				
4	0.3333	0.2500	0.2500	1.0000			
5	0.3333	0.2500	0.2500	0.2000	1.0000		
6	0.3333	0.2500	0.0000	0.5000	0.2000	1.0000	
7	0.3333	0.2500	0.0000	0.5000	0.2000	1.0000	1.0000

MIXED VARIABLE SETS: GOWER'S AND ANDERBERG'S COEFFICIENTS

Euclidean distance is ideal when measurements or ranks are involved, and could be used with presence/absence variables (as long as no distinction needs to be made between present–present matches and absent–absent matches). The Simple Matching Coefficient and Jaccard's Coefficient provide more elegant and simple ways of measuring similarity between cases when the dataset consists only of presence/absence variables. Neither works at all well with variables having more than two unranked categories, although, as noted above, such variables can be reformulated with each category as a separate presence/absence variable. There is also a different solution to the problem posed by such variables, as well as the problem posed by datasets consisting of several different kinds of variables. *Gower's Coefficient* and *Anderberg's Coefficient* have been devised for just such situations.

Gower's Coefficient between two cases is arrived at by calculating a score for each variable. The final coefficient of similarity is the mean of all the scores. The individual variable scores are arrived at differently for different kinds of variables:

- For a presence/absence variable, the Gower score is 1 for a present–present match and 0 for a mismatch. If there is an absent–absent match, the variable is omitted entirely (which is not the same as averaging in a score of 0). The treatment of presence/absence variables by Gower's Coefficient is thus like that of Jaccard's Coefficient.
- For a categorical variable whose categories are unranked the Gower score is 1 if the two cases belong to the same category and 0 if the two cases belong to different categories. Thus greater differences between numeric values assigned to the categories are ignored.
- For measurements and ranks, the absolute value of the difference between the values for the two cases is divided by the range of the measurements in the batch or, in the case of ranks, by the number of ranked categories the variable has. The quotient in either case is subtracted from 1 to produce the Gower score in the form of a similarity. This treatment is something like that provided by Euclidean distance for measurements and ranks.

A little experimentation with these rules for calculating the Gower scores will show that each score has a minimum value of 0 and a maximum value of 1. Thus the final coefficient (the average of the scores for all the variables) has a minimum value of 0 and a maximum value of 1. Expressed in this way it is also a similarity coefficient, not a dissimilarity coefficient, since larger values correspond to greater similarities.

Anderberg's Coefficient is closely related to Gower's and also involves the determination of scores for each variable that are averaged across all variables for each pair of cases:

- For a presence/absence variable, the Anderberg score is 1 for a mismatch or 0 for either a present–present match or an absent–absent match. It thus amounts, for presence/absence variables, to a kind of simple mismatching coefficient. That is, it works like the Simple Matching Coefficient turned into a dissimilarity where larger values indicate greater dissimilarity.
- For a variable with multiple unranked categories, the Anderberg score is 0 for a pair of cases falling in the same category or 1 for a pair of cases falling in different categories.
- For ranks, the Anderberg score is the absolute value of the difference between the category codes divided by one less than the number of categories. For example, if a variable has five categories (1, 2, 3, 4, and 5), cases coded 2 and 4, respectively, would receive a score of 2/4 or 0.5000.
- For measurements, the Anderberg score is the absolute value of the difference between the measurements for the two cases divided by the range of the measurements in the batch. Anderberg recommends using the square root of this score rather than the raw score to lessen the impact of outliers.

Once a score is determined for each variable, all the scores are averaged to produce the final Anderberg's Coefficient for the pair of cases under consideration. Like the Gower scores, the Anderberg scores have a minimum value of 0 and a maximum value of 1, so the final coefficient also ranges from 0 to 1. Unlike Gower's Coefficient, Anderberg's Coefficient, calculated in this way, is a dissimilarity coefficient. A value of 0 means identical cases; a value of 1 means totally dissimilar cases.

SIMILARITIES BETWEEN IXCAQUIXTLA HOUSEHOLD UNITS

Table 22.9 shows Gower's Coefficient of similarity between the household units at Ixcaquixtla from the data in Table 21.1. That dataset, as discussed in Chapter 21, contains both measurements and ranks, along with two presence/absence variables where present–present matches seem more meaningful than absent–absent matches. It was because of this mixture of variables for which different treatments seem appropriate that Gower's Coefficient was chosen. As a practical matter, it is always a good idea to examine a matrix of similarity scores like this. There are many possibilities for making mistakes – either with the software or in thinking through the principles of the chosen coefficient – and it is always reassuring to notice that pairs of cases whose values across the variables seem quite similar come out with high similarity scores, and the pairs of cases whose values across the variables seem quite different come out with low similarity scores. For example, Household Units 2 and 5 show up in Table 22.9 with a very high similarity score (0.8916). A look at Table 21.1 shows that these two household units have quite similar values on the majority of the variables. In contrast, Household Units 14 and 20 show up in Table 22.9 with a very low similarity score (0.3733). Again, a look at Table 21.1 is

Table 22.9. Gower's Coefficient of Similarity for the 20 Household Units from Ixcaquixtla, Based on the Data from Table 21.1

	1	2	3	4	5	6	7	8	9	10	11	12	13	14	15	16	17	18	19	20
1	1.0000																			
2	0.7198	1.0000																		
3	0.5120	0.4037	1.0000																	
4	0.8390	0.7036	0.6250	1.0000																
5	0.8191	0.8916	0.3936	0.6910	1.0000															
6	0.5864	0.4660	0.6970	0.7038	0.4548	1.0000														
7	0.8655	0.5853	0.6012	0.8209	0.6846	0.6855	1.0000													
8	0.7688	0.6889	0.4903	0.8316	0.6486	0.5597	0.6588	1.0000												
9	0.6589	0.9073	0.3629	0.6657	0.7989	0.4236	0.5245	0.6511	1.0000											
10	0.8794	0.7440	0.6006	0.9596	0.7314	0.6877	0.8154	0.8434	0.7028	1.0000										
11	0.5192	0.4159	0.6172	0.5790	0.4058	0.7202	0.4739	0.6031	0.3777	0.6004	1.0000									
12	0.7842	0.6125	0.6625	0.8794	0.6362	0.7425	0.8956	0.7110	0.5516	0.8488	0.4956	1.0000								
13	0.5757	0.5335	0.4539	0.5876	0.5224	0.6956	0.4562	0.6921	0.4999	0.6147	0.7905	0.4804	1.0000							
14	0.5084	0.6092	0.6351	0.4165	0.6532	0.3731	0.4009	0.4048	0.5751	0.4488	0.4695	0.3726	0.4338	1.0000						
15	0.7814	0.8696	0.4319	0.7389	0.8709	0.4974	0.6470	0.6687	0.8775	0.7793	0.4441	0.6741	0.5402	0.6104	1.0000					
16	0.5827	0.4623	0.8275	0.7170	0.4511	0.6048	0.6817	0.5673	0.4257	0.6898	0.5322	0.7745	0.3741	0.4808	0.4936	1.0000				
17	0.6737	0.5382	0.6269	0.8003	0.5257	0.6715	0.7574	0.7997	0.5004	0.7943	0.5587	0.8340	0.5592	0.2843	0.5735	0.7211	1.0000			
18	0.6779	0.4737	0.7931	0.5968	0.5531	0.5533	0.6364	0.5075	0.4250	0.6291	0.6498	0.5529	0.5160	0.7619	0.5230	0.6611	0.4645	1.0000		
19	0.7942	0.7228	0.6505	0.7915	0.7116	0.5689	0.6747	0.7291	0.6804	0.8274	0.6653	0.6989	0.5924	0.6042	0.7542	0.7395	0.6445	0.7077	1.0000	
20	0.8311	0.5509	0.5869	0.7621	0.6502	0.6463	0.9100	0.6493	0.4900	0.7594	0.4241	0.8550	0.4255	0.3733	0.6125	0.6457	0.7718	0.5892	0.6441	1.0000

Statpacks and Reporting Results

Measuring similarities between cases runs against the grain of the usual organization of statpacks, which are designed to work with variables rather than cases. Some statpacks nonetheless provide options for calculating similarity indexes between cases. Different statpacks vary quite substantially in how this is accomplished. In some, there is a specific set of options for measuring similarities between cases. In others, it is necessary to transpose the entire dataset so that rows (cases) become columns (variables) and columns become rows. Then the usual structure of measuring relationships between variables becomes a question of relationships between cases instead. In yet other statpacks, measuring similarities between cases is not a separate task but instead is embedded in the routines that perform a multivariate analysis that begins with similarities between cases. The statpack that will calculate the full variety of coefficients of similarity between cases discussed in this chapter is rare. There are, however, stand-alone specialty programs that perform just this task.

Whatever the software solution, in reporting the results of any multivariate analysis that starts with the measurement of similarities between cases, it is essential to specify how the similarities were measured. The choice of similarity coefficient has a major bearing on the outcome of a multivariate analysis, and the reader should always be made explicitly aware of that choice and the reasons for it.

---

consistent with this result, since Household Units 14 and 20 have quite different values for most of the variables. These calculations of Gower's Coefficient then jibe with the original data, reassuring us both that the choice of coefficient makes sense, and that it was calculated correctly.

Chapter 23
Multidimensional Scaling

Multidimensional scaling is perhaps in concept the simplest and most intuitive of the various approaches to multivariate analysis, and this can rightly be regarded as a major advantage. It is difficult to misunderstand the principles upon which it is based. A multidimensional scaling takes as its starting point a matrix of similarity (or dissimilarity) scores between cases like the one in Table 22.9. The analysis consists of an iterative, trial-and-error process of creating a configuration of points, each representing one of the cases in the dataset. These points representing the cases are placed in space in such a way that the rank order of the distances between the pairs of points corresponds as well as possible to the rank order of the similarity coefficients in space. That is to say, the aim of the configuration is to place the two points representing the two most similar cases closer to each other than any other pair of points in the configuration. The two points representing the second highest similarity score should be the second-closest pair of points, and so on. Finally, the two cases with the lowest similarity score should be the two points farthest apart in the configuration. In this very simple way, multidimensional scaling attempts to draw a picture of the relationships between cases that are encapsulated in the matrix of similarity coefficients. Since only a rank order correlation is sought between similarity scores and distances between pairs of points, multidimensional scaling is sometimes referred to as nonmetric multidimensional scaling.

The conceptual simplicity of multidimensional scaling masks the fiercely complex challenge of writing a program to produce such a configuration of points. A multidimensional scaling program must set up an initial configuration by placing points representing all the cases in space, and then tinker with that configuration, moving some points to new locations to see whether that improves the rank order correlation between distances between pairs of points and similarity

R.D. Drennan, *Statistics for Archaeologists*, Interdisciplinary Contributions
to Archaeology, DOI 10.1007/978-1-4419-0413-3_23,
© Springer Science+Business Media, LLC 2004, 2009

coefficients between pairs of cases. This is done over and over until no improvement can be found. As multidimensional scaling was developed, it was not unusual to get different results from different programs, but the algorithms for this iterative procedure are now honed enough that all the programs currently in use are pretty much equivalent. Some, but not all, large statpacks will perform multidimensional scaling.

It is easy to visualize multidimensional scaling in two dimensions, and even in three, but multidimensional scaling solutions can have more dimensions than physical space does. A perfect rank order correlation between similarity scores and distances between pairs of points can always be achieved in one less dimension than the number of variables. For the Ixcaquixtla household unit dataset, for example, which has ten variables, a configuration of points representing a perfect rank order correlation between similarity scores and distances between pairs of points can be achieved in nine dimensions. Since multidimensional scaling results are interpreted by looking at the configuration, however, this is an unsatisfactory solution. Looking at a configuration of points in nine dimensions is unbearably cumbersome – more difficult than simply looking at the original data table to hunt for patterns. The game, then, is to produce as good a rank order correspondence between similarity scores and distances between point pairs as possible in as few dimensions as possible. The smaller the number of dimensions, the easier it is to look at and interpret a multidimensional scaling configuration, so it is a great advantage to produce a configuration that represents the patterns in the similarity scores, not perfectly, but very accurately in very few dimensions. For any dataset, the larger the number of dimensions, the stronger the rank order correlation will be between distances between pairs of points and similarity scores between pairs of cases.

CONFIGURATIONS IN DIFFERENT NUMBERS OF DIMENSIONS

Carrying out multidimensional scaling starts by asking a statpack to take a set of similarity scores between cases (as described in Chapter 22) and produce the best possible configuration in one dimension. A one-dimensional configuration, of course, is an arrangement of points representing the cases along a line. Multidimensional scaling can be based on any one of several different rank order correlations, which are commonly referred to in this context as *stress* values. The different stress coefficients generally do not produce very different results. The lower the stress value, the better the rank order correlation between similarity scores and distances between pairs of points. For the matrix of similarity scores between Ixcaquixtla household units from Table 22.9, a one-dimensional configuration can be produced that has a final stress value of 0.3706. It is called a "final" stress value because the procedure is iterative, and a stress value is calculated at each step in the process. The iteration history of this scaling began with an initial configuration with a stress value of 0.4452. After the first successful iteration, which improved the configuration, the stress value was 0.4139. It then continued to drop, iteration by iteration, to 0.3988, 0.3874, 0.3795, 0.3709, 0.3707, and finally 0.3706. Beyond this, the algorithm could make no further improvement, and the analysis concluded. A stress value of 0.3706 is fairly high, and there was little intelligible patterning to be observed in the one-dimensional linear configuration.

The process is then repeated for two dimensions, which arranges points on a flat plane, easily represented as a scatter plot. For the Ixcaquixtla household dataset, the final stress value for the two-dimensional configuration is 0.1813, a substantial improvement over the 0.3706 of the one-dimensional configuration. We know that a three-dimensional configuration will enable an even better rank order correlation between similarity scores and distances between pairs of points. For the Ixcaquixtla data, the three-dimensional configuration yields a final stress value of 0.0726, another substantial improvement. There is further improvement in four dimensions, with the stress declining to 0.0465; and in five dimensions, with the stress down to 0.0332.

In practice, one must decide which configuration (the one-dimensional one, the two-dimensional one, the three-dimensional one, etc.) to attempt to interpret. Since interpretation centers on the examination of plots of the configuration of points, it is unusual to attempt to interpret a scaling in more than three dimensions. It simply becomes too cumbersome to attempt to visualize and inspect a configuration of points in more dimensions than actual physical space provides. A two-dimensional configuration is easier to inspect than a three-dimensional one, and a one-dimensional configuration is easier still to inspect. A one-dimensional configuration will not, however, be at all easy to interpret if it does not provide a reasonably accurate picture of the pattern of relationships between cases encapsulated

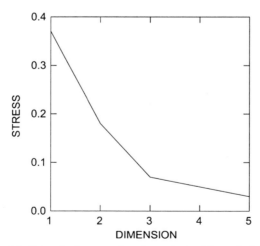

Figure 23.1. Graph of declining final stress values for analysis of Ixcaquixtla household data with increasing number of dimensions.

in the matrix of similarity scores. It is the stress value that indicates how accurate a picture it is.

Sometimes it is helpful to look at a plot of declining stress values like the one in Fig. 23.1. This shows the high stress value for one dimension mentioned above, the much lower stress value for two dimensions, and an additional substantial decline in stress for the three-dimensional configuration. Beyond this, in four and five dimensions, the stress continues to decline, as it always will, but at a much slower pace. This "elbow" in the graph of declining stress value is an indication that the three-dimensional configuration may be a good representation of the patterns in the dataset, and that, since there is less improvement in four and five dimensions, there may be little to be gained in looking at these configurations. There is also a useful rule of thumb that stress values of about 0.1500 or less are often associated with interpretable configurations. For the Ixcaquixtla household scaling, it is the three-dimensional configuration that breaks through to a stress value below 0.1500. For this reason, then, and because of the elbow in the graph at three dimensions, it seems likely that the three-dimensional configuration will be an effective representation of the patterning in the dataset.

Table 23.1. Coordinates in Three Dimensions of the Multidimensional Scaling of Household Units from Ixcaquixtla

Household Unit	Dimension 1	Dimension 2	Dimension 3
1	−0.285	0.301	0.142
2	−1.069	−0.003	−0.083
3	0.996	−0.568	0.523
4	0.062	0.421	−0.088
5	−0.963	0.011	0.224
6	0.970	0.056	−0.517
7	0.178	0.619	0.427
8	−0.146	0.425	−0.624
9	−1.228	−0.008	−0.159
10	−0.059	0.348	−0.105
11	0.739	−0.761	−0.830
12	0.314	0.626	0.216
13	0.102	−0.466	−1.261
14	−0.541	−1.384	0.527
15	−0.881	0.064	0.024
16	0.817	0.017	0.716
17	0.504	0.780	−0.242
18	0.382	−0.935	0.504
19	−0.121	−0.305	0.082
20	0.229	0.762	0.524

INTERPRETING THE CONFIGURATION

The essential element in the results of this analysis, as always with multidimensional scaling, is the list of coordinates for each case in the three dimensions (Table 23.1). Inspecting a multidimensional scaling configuration usually means plotting the points on a graph to examine their relationships. For some datasets, simply plotting the points and labeling them makes the patterning clear, as seen in Fig. 23.2. This two-dimensional scaling configuration was produced from a matrix of Euclidean distances in real space – the matrix of distances between each pair of cities. With only this information, a multidimensional scaling analysis placed the cities in a two-dimensional configuration that represents their actual physical locations quite accurately. Simply labeling the points in a configuration like this makes the nature of the patterning obvious.

Interpretation of the multidimensional scaling of the Ixcaquixtla household units is considerably more complicated. In the first place, the configuration we need to interpret is not in two dimensions, but in three. It is usually easiest to see three-dimensional configurations in three views based on each pair of the three dimensions taken two at a time. If the three-dimensional configuration is imagined in real space, it would take the form of a cube with points scattered around in it. The cube could be looked at in perspective from an angle, and most statpacks will produce such a plot, but it can be very difficult to see just where the points really are in relation to each other. It is often clearer to look at the cube successively from three different sides: first, directly from the front; second, directly from one side; and third, directly from the top. This is how we will inspect the three-dimensional configuration produced from the Ixcaquixtla household data.

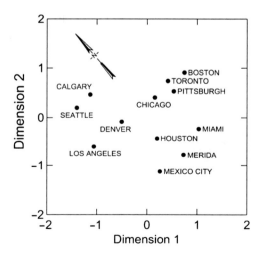

Figure 23.2. Two-dimensional scaling configuration based on distances between cities.

A second complexity in this analysis is that labeling the household units with their numbers would not automatically make any patterning clear. The simple act of labeling the cities in Fig. 23.2 makes the patterning clear because we know where they are, and we immediately recognize that they have been placed in their actual spatial locations. We have no prior knowledge of, say, Household Unit 6 that makes such pattern recognition possible. In a case like this, the most useful strategy is likely to be looking at the behavior of each variable, one at a time, in the space defined by the three-dimensional configuration.

Figure 23.3 begins this process. The cube of the three-dimensional configuration is looked at in the form of three plots. The first looks directly along the third dimension, to give a clear orthogonal view of the configuration of points in Dimensions 1 and 2. The second takes a view directly along the second dimension, showing the configuration in Dimensions 1 and 3. And the third takes a view directly along the first dimension, giving a clear view of Dimensions 2 and 3. To envision the full three-dimensional configuration as a cube, imagine cutting out the leftmost plot in Fig. 23.3 and pasting it to the top of the cube. Then cut out the center plot and paste it to the left side of the cube. The rightmost plot would go on the front of the cube. Together the three plots make it possible to look at the cube in all its dimensions. Within the plots in Fig. 23.3, each circle represents one household, and larger circles correspond to higher values of the first variable in the dataset, Bowls as % of Sherds. A clear trend is visible in the plot of Dimensions 1 and 2. The values of Bowls as % of Sherds are quite low in the upper right corner of this plot and increase steadily toward the lower left. Household units with high proportions of bowls, then, appear toward the lower left of the plot of Dimensions 1 and 2.

Figure 23.4 provides the same kind of illustration of Energy Invested in Burials, and again we see the same pattern. Households with the highest levels of energy invested in burials on average appear toward the lower left corner of the plot of Dimensions 1 and 2. Continuing with Fig. 23.5, we see a pattern for Decoration as % of Sherds that is not identical, but very similar to that seen for the two variables shown in Figs. 23.3 and 23.4. Household units with the highest proportions of decorated ceramics appear toward the bottom and slightly toward the left in the plot of Dimensions 1 and 2. The Fauna/Sherd Ratio shows the same pattern yet

Figure 23.3. Plots of the three-dimensional scaling of Ixcaquixtla household data (larger circles indicate higher proportions of bowls).

Figure 23.4. Plots of the three-dimensional scaling of Ixcaquixtla household data (larger circles indicate greater energy investment in burials).

Figure 23.5. Plots of the three-dimensional scaling of Ixcaquixtla household data (larger circles indicate higher proportions of decorated ceramics).

Figure 23.6. Plots of the three-dimensional scaling of Ixcaquixtla household data (larger circles indicate higher proportions of faunal remains).

again: higher values toward the bottom and slightly toward the left in the plot of Dimensions 1 and 2 in Fig. 23.6.

These four variables, then, pattern in the same way in the space defined by the three-dimensional configuration. Households where the value of one of these variables is high tend strongly to be households where the values of the others are also high. The four variables form a gradient running roughly from the upper right to the lower left in the plot of Dimensions 1 and 2. Values for proportion of bowls, for energy investment in burials, for proportion of decorated ceramics, and for the ratio of faunal remains to sherds increase in a more or less gradual fashion along this gradient. None of the other variables, as we shall see, pattern in this way. The description up to this point is of the patterning to be observed in the dataset, and it is clearly shown in the plots of these variables in the multidimensional scaling space. The next step takes us from the realm of finding patterning through multivariate analysis into the realm of interpretation. We might interpret this pattern as reflecting a dimension of economic differentiation at Ixcaquixtla. All four variables might plausibly be connected to economic well-being or standard of living. All four pattern in the same way, suggesting a gradient of wealth from low to high.

Moving on through the variables, Platform also shows a clear pattern in the plot of Dimensions 1 and 2 in Fig. 23.7, but it is a distinctly different pattern from the one we have been seeing. High values for Platform are clustered toward the upper left. Since Platform is a presence/absence variable, high values (large circles) mean the code for presence (1) and low values (small circles) mean the code for absence (0). The other presence/absence variable, Mace Heads, shows a very similar pattern in Fig. 23.8. Again as the inevitable result of being a presence/absence variable the pattern looks more like a cluster than a gradient. Closer comparison of Figs. 23.7 and

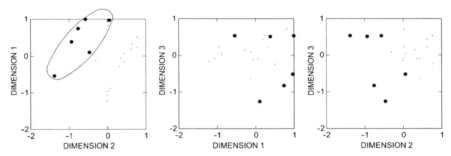

Figure 23.7. Plots of the three-dimensional scaling of Ixcaquixtla household data (larger circles indicate house structures built on platforms).

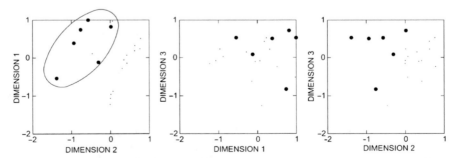

Figure 23.8. Plots of the three-dimensional scaling of Ixcaquixtla household data (larger circles indicate presence of maces in burials).

Figure 23.9. Plots of the three-dimensional scaling of Ixcaquixtla household data (larger circles indicate higher proportions of marine shell).

23.8 shows that this pattern is actually a bit more like a gradient than it first appears. Farthest off to the upper left are four household units where the house structure is built on a platform and where mace heads are present in burials. Not so far toward the upper left are two households with platforms but no mace heads and two with mace heads but no platforms. At the opposite end of the scale is a larger number of households with neither platforms nor mace heads. These are toward the lower right in the plot of Dimensions 1 and 2.

There is, then, another gradient running perpendicular to the one apparent in Figs. 23.3–23.6. Since they are perpendicular, the two are unrelated to each other. Some households with platforms and mace heads are toward the wealthier end of the first gradient; others are not. In this way Figs. 23.7 and 23.8 delineate a second separate and independent element of patterning in the multidimensional space. The patterning is again quite clear, although as always open to potentially different interpretations. We might interpret this second gradient as one of prestige or possibly political authority, which at Ixcaquixtla seems not to correspond to wealth. However we interpret these two gradients in the multidimensional space, both their presence and their independence are clear patterns in the scaling results.

The patterns we have discussed up to this point have been most visible in the plot of Dimensions 1 and 2. We have not yet needed to look at the cube from any other angle. A few household units with high proportions of marine shell, however, can be seen to cluster clearly together in the plot of Dimensions 2 and 3 in Fig. 23.9. They also form a detectable cluster in the plot of Dimensions 1 and 2, although this cluster is less clear, since it also includes some households without high proportions of shell. This pattern simultaneously suggests both some relation with the gradient identified in Figs. 23.7 and 23.8 and some independence from it. A few households with high proportions of obsidian also cluster together in the plot of Dimensions 2 and 3 in Fig. 23.10, but this cluster is not in the same place as the cluster of household units with high proportions of shell. The cluster of households with high proportions of obsidian can also be observed in the plot of Dimensions 1 and 3. Households with high proportions of obsidian also focus in the upper left of the plot of Dimensions 1 and 2, but they are more mixed there with households having low proportions of obsidian. If high proportions of marine shell and obsid-

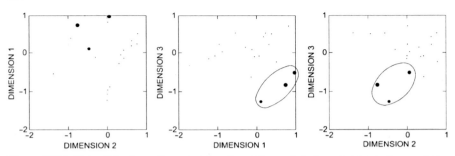

Figure 23.10. Plots of the three-dimensional scaling of Ixcaquixtla household data (larger circles indicate higher proportions of obsidian).

Figure 23.11. Plots of the three-dimensional scaling of Ixcaquixtla household data (larger circles indicate higher proportions of kiln wasters).

ian are interpreted to reflect stronger contacts with the distant regions from which they come, then there does not seem to be much correspondence between households connected to the obsidian source regions and those connected to the coasts. Nor do either of these clusters correspond in any way to the gradient interpreted as wealth, but both the shell and obsidian clusters do pattern generally toward one end of the gradient interpreted as prestige. Taken together these interpretations would suggest that, for the inhabitants of Ixcaquixtla, contacts with distant regions were separately maintained by different households and had little to do with wealth. They do, however, show some sort of relationship to what was interpreted as prestige.

Fig. 23.11 shows a cluster of household units with high proportions of kiln wasters, a cluster that is especially clear in the plot of Dimensions 1 and 3. This view of the configuration also reveals a cluster of household units with high proportions of lithic debitage (Fig. 23.12). The two clusters are located in different places in the multidimensional scaling space, showing us that they are two different sets of households with high proportions of these artifact types. Neither high proportions of debitage nor high proportions of kiln wasters align with either of the gradients observed in the Dimensions 1 and 2 view. In the plot of Dimensions 1 and 3, the kiln waster cluster overlaps with a manifestation of the obsidian cluster, raising the possibility of a relationship between these two.

Figure 23.12. Plots of the three-dimensional scaling of Ixcaquixtla household data (larger circles indicate higher proportions of lithic debitage).

Figure 23.13. Plots of the three-dimensional scaling of Ixcaquixtla household data with patterning and possible interpretations indicated (household units indicated by number).

Fig. 23.13 summarizes the patterns seen in the three-dimensional scaling configuration, labeled with some of the plausible interpretations. To re-emphasize, the multidimensional scaling analysis has not shown us that these interpretations are correct. It has, however, shown us that there is a set of characteristics (large amounts of energy invested in burials and high proportions of bowls, decorated ceramics, and faunal remains) that go together in this dataset on household units at Ixcaquixtla. They parallel each other sufficiently strongly to identify a gradient across the configuration when it is looked at from a particular angle (Fig. 23.13 left). The nature of this pattern is also better described as a gradient across the space than as a distinct cluster of household units with very high values for these four variables. It represents not a sharp division between household units that have these characteristics and those that don't, but a more gradual range of variation. If we interpret this pattern as connected to wealth distribution, then we have learned something about the nature of wealth distribution at Ixcaquixtla.

The multidimensional scaling has also shown us a gradient relating houses on platforms to mace heads in burials (Fig. 23.13 left). These two characteristics tend to go together and also appear to form more of a gradient than a sharply distinguished cluster. This gradient is entirely unrelated to the one discussed in the previous paragraph. The multidimensional scaling does not prove that this gradient is related to prestige, but it shows that the gradient exists. If we interpret it as prestige, then we

Statpacks and Reporting Results

Multidimensional scaling was once a specialized domain of its own with several major programs dedicated specifically to performing it. By now, many large multipurpose statpacks perform multidimensional scaling, but not all of them have incorporated it into their repertoire. Sometimes the process of measuring similarities between cases is incorporated into the multidimensional scaling routines themselves; sometimes it is treated as a separate task. Whichever way it is accomplished, the measurement of similarities between cases is conceptually a task separate from scaling. As noted in Chapter 22, it is vitally important in reporting the results of a multidimensional scaling analysis to specify what coefficient was used to measure the similarities between cases. The choice of a measure of similarity is the single choice made in performing the analysis that is likely to have the biggest impact on the outcome. Readers deserve to know exactly what choice was made (and exactly what the variables were) so that they can judge for themselves how appropriate the choice was.

Once the similarities have been measured, performing the scaling is likely to involve running the analysis several times, first in one dimension, then in two dimensions, and so on. This is the only way to obtain the final stress values for configurations in different numbers of dimensions so as to be able to decide how many dimensions to work with. Looking for stress values below 0.1500 and looking for an "elbow" in the plot of final stress values against number of dimensions are useful indicators of how many dimensions are needed. In the final analysis, however, the most compelling reason to select a particular number of dimensions is that that configuration shows clear and sensible patterning.

The central element in statpack output for a multidimensional scaling analysis is the list of coordinates of the points corresponding to the cases. There will be a coordinate for each point in however many dimensions are selected. Statpacks usually provide for saving this list of coordinates in multidimensional space as a data file which can then be combined with the original variables so as to use the statpack's tools for making scatter plots to produce plots of the configuration. In these plots, the points may be labeled to show which case is which, if such visual identification is meaningful. Such labeling would not have been much help in interpreting the Ixcaquixtla household scaling, and we relied instead on plots in which symbols of different sizes indicated the values of each variable in turn. Most statpacks have an option to vary the symbol size according to the value of some variable in the dataset. An essential element in the presentation of results is one or more plots of the configuration so that readers can see the patterns of points that you are interpreting.

It is a good idea to limit the number of variables in a multidimensional scaling analysis to no more than about half the number of cases. If the number of variables is much larger than this, there is substantial risk of finding spurious patterns that are no more than the product of random noise in the data.

have learned something about the nature of prestige at Ixcaquixtla and about its lack of relation to wealth distribution.

The multidimensional scaling has also shown us two different clusters of households with high proportions of material brought from long distances away (obsidian and shell, Fig. 23.13 center and right). Both are more accurately characterized as clusters, sharply set off from other household units, than as gradients of continuous variation. And the two do not overlap where they are seen most clearly. Both these clusters are, however, seen somewhat more hazily in the zone interpreted in the plot of Dimensions 1 and 2 as highest in prestige. In similar fashion, the multidimensional scaling has shown us two different clusters of households with high proportions of artifacts likely related to craft production (kiln wasters and lithic debitage, Fig. 23.13 center). These are also more accurately characterized as clusters than as gradients, and these also do not overlap with each other or correspond to the gradients in the plot of Dimensions 1 and 2. The kiln waster cluster, though, is seen to be largely coterminous with the obsidian cluster as it appears in the plot of Dimensions 1 and 3; these two things do coincide in some households.

All this could have turned out differently. The multidimensional scaling might, for example, have shown us that a cluster of households with high proportions of lithic debitage corresponded well to a cluster of households with high proportions of obsidian. This might have led us to think of some degree of special focus in these households on various aspects of lithic raw material procurement and production combined. The patterns we actually see, in contrast, lead us to think of these two as special activity foci, but not combined in the same households.

Multidimensional scaling has been quite successful at drawing us a picture of patterns in the variation that exists in the multivariate dataset on household units at Ixcaquixtla. That picture has led us to a series of observations and rather complicated characterizations of those patterns. Clusters and gradients are two common kinds of patterns to be identified in multidimensional scaling configurations, but many other sorts of spatial patterns are imaginable. What it is possible to perceive in a multidimensional scaling configuration is up to the imagination of the analyst. This is simultaneously a major advantage and a disadvantage of multidimensional scaling. Identifying patterns in scaling results is neither automatic nor necessarily simple. A good bit of time is likely to be consumed in producing and inspecting various kinds of plots of configurations. Finally, though, there is the possibility of observing a richly varied array of patterning. There is also the possibility of observing very little in the way of patterning in a scaling configuration. This may at first seem a disadvantage, but it is really quite a substantial advantage. If little or no meaningful patterning shows up in a scaling configuration, it may well reflect a general lack of meaningful patterning in the dataset. This is not a happy outcome, but it is one of the possibilities in the real world. If there is indeed little meaningful patterning in a multivariate dataset, we do want an analytical approach that can tell us that.

Chapter 24
Principal Components Analysis

In moving from multidimensional scaling to principal components analysis, we shift from the simplest and most commonsense approach conceptually to the most abstract and mathematical. Mastering the mathematical fundamentals of principal components analysis is a lot of work – work that does not finally bring much payback in making it easier to perform more reliable or successful analyses. In keeping with the overall approach of this book, we will give short shrift to the abstract mathematical fundamentals of principal components analysis and concentrate our attention on understanding its principles and concepts in ways that provide surer guides to effective use of the technique. This approach is very different from the one that is usually taken to the subject. Nevertheless, more archaeologists seem able to understand the principles of principal components analysis more readily, more deeply, and to better effect through such a commonsense approach than through an abstract mathematical explanation. Understanding and effective use of multidimensional scaling does not require much knowledge of how the iterative trial-and-error procedure that produces the configuration is programmed. In similar fashion, what principal components are and how they tell us about patterning in a multivariate dataset can be understood effectively without much knowledge of the particular mathematics that produce them.

Principal components analysis is often confused with factor analysis. Opinion is divided about how much this confusion matters. There certainly are distinctions between the underlying logic of the two analytical techniques. On the other hand, their results are presented and interpreted in precisely the same way. At the practical level, it is extremely unusual to carry out the two analyses on the same data and get very different results. Not surprisingly, statpacks tend to have a focus on the practical, and principal components analysis and factor analysis are often combined into one set of routines where the choice between the two is simply one of the options to set. The difference between them certainly matters little to the commonsense approach of this chapter. The vocabulary we will use will be that of

R.D. Drennan, *Statistics for Archaeologists*, Interdisciplinary Contributions to Archaeology, DOI 10.1007/978-1-4419-0413-3_24,
© Springer Science+Business Media, LLC 2004, 2009

principal components analysis, but in actual fact, this chapter could just as easily be a chapter on factor analysis. Virtually the only difference would be to replace the words "principal component" or "component" with "factor."

Although we did not discuss it in exactly this way, it would have been easy enough to characterize multidimensional scaling as a way of reducing the ten variables measured or categorized for the Ixcaquixtla household units to three variables (the three dimensions of the scaling configuration we found patterns in). In some sense, the major elements of the patterning to be found in Table 21.1 were encapsulated in the simpler and more compact Table 23.1. This is much more centrally and directly the idea behind principal components analysis, which can be looked at as a way of reducing a large number of variables to a much smaller number of variables that still reflects reasonably accurately (although not perfectly) the major patterns in the original dataset.

Multidimensional scaling's effort to produce as good a configuration as possible in as few dimensions as possible bears more than a passing resemblance to principal components' effort to reduce the number of variables as much as possible without losing important aspects of the patterning in the original dataset. The approach taken by principal components analysis to this task, however, does not begin by measuring the similarities between all pairs of cases. It begins instead by looking at relationships between variables. Usually this is done with tools we used in Chapter 15. The point of departure for principal components analysis is a matrix of correlations between all pairs of variables in the original dataset. This matrix tells us the same kind of thing about relationships between variables that the matrix of similarity scores we used for multidimensional scaling tells us about relationships between cases. If two variables show a strong correlation, that means they behave quite similarly (have high values for the same cases and low values for the same other cases).

CORRELATIONS AND VARIABLES

The broad thought behind principal components analysis is that a set of variables that all show strong correlations with each other are all responding to the same underlying thing and that these variables could, in some sense, be replaced in the dataset by a single variable with little damage to the overall patterning of relationships between cases or variables that characterizes the original dataset. The dataset would thus, in some sense, be re-expressed with fewer variables. As far as the user of principal components analysis is concerned, there might well be some iterative trial-and-error procedure by which such a task is accomplished, much like multidimensional scaling. This is not, in fact, how the trick is done in principal components analysis though. Principal components are extracted mathematically by working with the matrix of correlations between variables. The goal is to produce a set consisting of as few components as possible that show strong correlations with the original variables.

The fact that principal components analysis starts with correlation coefficients is important. As we saw in Chapter 22, a number of different similarity coefficients have been devised for dealing with similarities between cases with different sorts of variables. Correlation and regression, as we saw in Chapter 15, is built on scatter plot logic and most suitable for measurements. If all the variables in a multivariate dataset are measurements, then looking at the relationships between them by way of correlation coefficients makes sense. It makes less sense if some of the variables are ranks or categories. In practice, principal components analysis often does produce sensible and valid results even when the variable set does not consist purely of measurements. It should not be too surprising that variables that are ranks rather than true measurements are not especially threatening to principal components analysis. As we saw in Chapter 16, rank order correlation coefficients are a better tool for relating ranks than regression and correlation, but a correlation coefficient (r) gives a decent approximate assessment of the degree of correlation between variables that are ranks.

Unranked categories are a different proposition. The scatter plot logic of regression and correlation means that values of 1 and 3 are treated not only as more different than values of 1 and 2, but also as twice as different. (The difference between 1 and 3 is 2, and the difference between 1 and 2 is 1.) We faced a very similar problem in thinking about Euclidean distance in Chapter 22. We can consider, as we did before, the possibility that the Ixcaquixtla household dataset had a variable for type of wall construction, and that the categories were wattle-and-daub, wood-plank, and mud-brick, assigned values of 1, 2, and 3, respectively. It does not seem at all reasonable to treat 1 and 3 as any more different than 1 and 2, but correlation coefficients (like Euclidean distances) inevitably will do this. This kind of category variable with multiple unranked categories is truly unsuitable for measurement of relationships with other variables by way of correlation coefficients and thus is truly unsuitable for principal components analysis. We came to the same conclusion about Euclidean distances, and the same solution discussed there is potentially applicable in principal components analysis. The three categories of kinds of wall construction can be reorganized into three separate presence/absence variables.

Category variables with two categories (including presence/absence variables), of course, are also not the most suitable fodder for regression and correlation. If the question is simply to assess the strength and significance of the relationship between a two-category variable and some other variable, we would not choose regression and correlation. Principal components analysis, however, must begin with correlations, and it turns out that correlations, while providing only a blunt instrument for assessing the strength of relationships involving two category variables, can provide an acceptable rough approximation.

Imagine the scatter plot we would draw to explore the relationship between two presence/absence variables. Since the values of each of these two variables would be limited to 0 and 1, there are only four places in a scatter plot where points could fall: where $x = 0$ and $y = 0$ (the origin of the graph at its lower left corner), where $x = 1$ and $y = 1$ (the upper right corner), where $x = 1$ and $y = 0$ (the lower right corner), and where $x = 0$ and $y = 1$ (the upper left corner). If the two variables are strongly related positively, that means that when *x* is 1, *y* tends also to be 1, and that when *x* is 0, *y* tends also to be 0, and most of the points will fall at the lower left corner and the upper right corner. The best-fit straight line in the scatterplot will run from the lower left corner to the upper right corner, the correlation will be positive, and if there are very few points at the other two corners, the correlation will be fairly near 1. If the two variables are strongly related negatively, the same thing will happen, but the line will run from the upper left corner to the lower right corner for a correlation coefficient near -1. If the two variables are not strongly related, then points will be broadly distributed across all four possible locations, the best fit straight line will not be a very good fit, and the correlation coefficient will be closer to 0.

In sum, the result of measuring the strength of relationships between presence/absence variables with a correlation coefficient is crude but functional – functional enough to make it possible to use presence/absence (or other two-category) variables in a principal components analysis. Correlation coefficients do a better job with ranks, and, of course, they are just the right tool for real measurements. To reiterate, the one kind of variable that simply must be gotten out of a dataset before principal components analysis is a variable with multiple unranked categories.

EXTRACTING COMPONENTS

The procedure for extracting principal components can be thought of as a multidimensional equivalent of finding best-fit straight lines. If there is a perfect correlation between two variables, we know that all the points in the scatter plot lie exactly on the best-fit straight line. In such a situation, the two variables that form the axes of the coordinate system of the scatter plot could be done away with and replaced by a single axis running along the best-fit straight line. Coordinates along this single axis would enable us to position the points perfectly in the scatterplot, and two dimensions of variability would have been re-expressed or reduced to one. If the correlation between the two original variables is strong but not perfect, we could reduce the two axes of the scatterplot to one running along the best-fit straight line, and reproduce the pattern of the scatter plot, not perfectly, but pretty well. If the correlation between the two original variables is quite weak, then reducing the scatter plot to coordinates along a single axis would do quite a poor job of capturing the pattern of the points in the scatter plot.

Principal components analysis can be visualized as beginning with a scatter plot in as many dimensions as there are variables in the initial dataset. Something akin to a single best-fit straight line is determined for this multidimensional scatter plot, and this becomes the first component. This component will align relatively closely with one or more of the original variables, which is the same as saying that it will show a strong correlation with one or more of the original variables. To the extent that several of the original variables are strongly correlated with each other, then this first component can simultaneously show strong correlations with all of them.

Since this first component is akin to a best-fit straight line, it can be thought of as accounting for as much variability as possible and leaving residuals. The process is repeated for extracting a second component which accounts for as much as possible of the variation in the residuals left by the first, again by leaving residuals that are as small as possible. The analysis continues, extracting a third component, then a fourth, and so on.

Components are described in terms of their correlations with each of the original variables. These correlations are usually referred to as *component loadings*. Since it was in some sense created to minimize residuals, the first component will have fairly large loadings on a fairly large number of the original variables. Since the loadings are correlation coefficients (r) between the component and each of the original variables in turn, their squares (r^2) express the proportion of variation in each of the original variables explained by the component. The squared loadings of all the original variables on a component are often summed up, and this sum of the squared loadings (the sum of the component's r^2 values with the original variables) is called the *eigenvalue*. Since the eigenvalue is the sum of the proportions of variation explained for each of the variables in turn, the eigenvalue divided by the number of variables is the overall proportion of variation in the original dataset explained by a component.

There is an eigenvalue for each component. The eigenvalue of the first component is the highest; of the second component, the second highest; and so on. If the number of components extracted is the same as the original number of variables, then all the variation in the original dataset is always explained. If all the eigenvalues are divided by the number of variables so as to express what proportion of the overall variation is explained by each, these eigenvalues, each divided by the number of variables, will sum up to 1, reflecting the fact the all the components taken together explain 100% of the variation in the original variables. If many of the variables are strongly correlated with each other, then the first few components will be able to account for a very large proportion of the variation in the original dataset. Their eigenvalues will be relatively large, and the eigenvalues of the last components will be quite small. Components with large eigenvalues are the most meaningful, encompassing as they do the largest proportion of the variation in the dataset.

CARRYING OUT THE ANALYSIS

The implications of this entire line of thinking become clearer when it is put into practice. Seven of the variables for household units at Ixcaquixtla are measurements, one is a set of ranks, and two are presence/absence variables (Table 21.1). They are thus not perfectly qualified for correlation analysis, but represent the kind of real-world compromise with perfection that is often made in order to carry out a principal components analysis. The extraction of ten components produces the set of eigenvalues in Table 24.1. The sum of the eigenvalues divided by the number of

**Table 24.1. Eigenvalues for Principal Components
Extracted from the Ixcaquixtla Household Data**

Component	Eigenvalue	Eigenvalue/No. of Variables
1	3.511	0.3511
2	2.291	0.2291
3	2.100	0.2100
4	0.887	0.0887
5	0.473	0.0473
6	0.326	0.0326
7	0.213	0.0213
8	0.110	0.0110
9	0.063	0.0063
10	0.027	0.0027

**Table 24.2. Component Loadings (Unrotated) for Analysis of the
Ixcaquixtla Household Dataset**

	Components				
	1	2	3	4	5
Energy in Burials	−**0.944**	0.173	0.052	0.063	0.054
Fauna/Sherds	−**0.933**	0.197	0.017	0.046	0.041
Bowls %	−**0.909**	0.223	0.007	0.193	0.145
Decoration %	−**0.858**	0.041	0.193	0.070	0.261
Platform	0.205	**0.905**	0.111	0.251	0.067
Mace Heads	0.207	**0.750**	0.399	0.285	0.031
Shell/Sherds	0.108	**0.683**	0.640	0.171	0.118
Wasters %	0.157	0.291	**0.788**	0.032	0.481
Obsidian %	0.253	0.479	**0.710**	0.327	0.257
Debitage %	−0.051	0.086	0.593	**0.747**	0.249

variables is 1, just as it should be, indicating that these ten components together account for 100% of the variation in the household data.

As the eigenvalues in Table 24.1 make clear, each of the first three components explains considerably more variation than subsequent ones do. Taken together these three explain 79% of the variation in the dataset, and the first three components will probably convey much or all of the meaning to be found in these results. Since it is often useful to look a bit beyond such probable limits, the loadings for five components are given in Table 24.2.

Four variables have very high loadings on the first component: energy invested in burials, the ratio of faunal remains to sherds, the proportion of bowls among the ceramics, and the proportion of decorated sherds. These are the four variables that the multidimensional scaling put together in parallel, forming a gradient across the plot of Dimensions 1 and 2. Both analyses, then, show us this same element of patterning in the dataset, which it was suggested in Chapter 23 might be interpreted as wealth. It is of no consequence that the signs of the component loadings for all four of these variables are negative. The important observation is that the loadings are strong and the signs of all four are the same. If energy invested in burials had a strong negative loading and the fauna/sherd ratio had a strong positive loading, this would mean that high energy investment in burials and low fauna/sherd ratios corresponded to this component (and thus to each other). Since all the signs are the same, we know that high energy investment in burials, high fauna/sherd rations, high proportions of bowls, and high proportions of decorated ceramics all correlate with each other on this component. None of the other variables show loadings of much strength at all on this component, suggesting that none of them relates very strongly to wealth (or whatever it is that this element of patterning in the dataset may represent).

Three variables have strong loadings on the second component (again all with the same sign): presence of a platform, presence of mace heads in burials, and the shell/sherd ratio. We saw a similar relationship between platforms and mace heads in the multidimensional scaling, and we also noted there a slightly more ambiguous relationship between these two things and a high ratio of shell. The shell/sherd ratio loads strongly on the second component here but its divided loyalties, so to speak, are reflected in its loading on the third component, which is almost as strong. The four variables that have such high loadings on the first component have very low loadings on the second, and the three variables with high loadings on the second component have very low loadings on the first. The message for us in this observation is that the first two components are quite independent of each other. We drew this same conclusion from the fact that these two elements appeared in the multidimensional scaling as two gradients that were perpendicular to each other.

The two variables with the strongest loadings on the third component are the proportions of kiln wasters and obsidian. We saw that high proportions of these two things formed clusters that overlapped in one view of the multidimensional scaling configuration, but not so much in another view in which they also appeared. The shell/sherd ratio also loads fairly highly on this third component. It is interesting to note that obsidian also has a moderate loading on the second component, paralleling the moderate relationship we saw in the multidimensional scaling between high proportions of obsidian and the gradient interpreted there as prestige. Finally, though, all four variables that did not have extremely high loadings on the first two components, have moderate to strong loadings on the third component. In part, this is produced by the fact that household units with high values for these four variables tend to share lower values for the six variables that have very high loadings on the first two components.

The only loading of much strength on the fourth component is for the proportion of debitage. It is in this way that this variable's particular lack of connection to others is shown in the principal components results. In the multidimensional scaling, both obsidian and kiln wasters stood partly apart and partly together, and both obsidian and shell were connected in their tendency to be at the high end of the gradient interpreted as prestige. Debitage, however, formed a cluster with little indication of overlap or relationship to anything else in the multidimensional scaling, and it stands apart in the principal components analysis as well.

**Table 24.3. Component Loadings (with Orthogonal Rotation) for Analysis of the
Ixcaquixtla Household Dataset**

	Components				
	1	2	3	4	5
Energy in Burials	**−0.960**	−0.008	−0.044	0.068	0.047
Fauna/Sherds	**−0.950**	0.045	−0.061	0.020	0.068
Bowls %	**−0.937**	−0.027	0.129	−0.127	0.151
Decoration %	**−0.854**	−0.128	−0.164	0.022	−0.272
Shell/Sherds	0.002	**0.946**	−0.058	−0.151	0.103
Mace Heads	0.068	**0.908**	0.099	0.087	0.023
Platform	−0.008	**0.597**	**0.743**	−0.087	−0.156
Obsidian %	0.089	−0.088	**0.934**	0.166	−0.238
Debitage %	0.004	0.051	−0.097	**−0.970**	0.169
Wasters %	0.041	−0.086	0.374	0.224	**−0.874**

There is considerable difference of opinion about *rotation* of loadings in principal components analysis. Once the components have been extracted, they can be
thought of as coordinate axes in a space of multiple dimensions. This whole set
of extracted axes can be rotated around in the space to maximize various different criteria of relationships with the original variables. They can be subjected to
orthogonal rotation in which the axes are rotated as a set and all are kept at right
angles to each other (i.e., they are uncorrelated with each other). Or they can be
rotated individually in *oblique* rotation, in which they lose the property of being
uncorrelated with each other. Within each of these two large families of rotation
there are several variants.

Orthogonally rotated component loadings for the Ixcaquixtla household analysis
are given in Table 24.3. As usual with orthogonal rotation the contrasts between high
loadings and low loadings on each component are maximized. The same four variables that showed strong loadings with the same sign on the first component in the
unrotated components behave the same in the rotated components. The same three
variables load strongly on the second component in both sets as well. In the rotated
components obsidian appears more strongly linked to platforms by virtue of their
strong loadings on the third component. The relationship between obsidian and kiln
wasters that we saw in the unrotated components and in the multidimensional scaling has disappeared, and both debitage and kiln wasters stand strongly apart from
the other variables. The two sets of component loadings correspond quite well in
regard to the main elements of patterning, and sometimes these main elements show
up more sharply in rotated component loadings. The discrepancies in the indications of some of the lesser elements in the patterning exemplify what worries some
analysts about component rotation.

There is no easy answer, and certainly no consensus answer, to the question about
the wisdom of rotating components. Large statpacks all do principal components
and factor analysis, and they all offer rotation as an option. There is certainly no
harm in looking at both rotated and unrotated components. If the important elements of patterning in both sets of loadings correspond well to each other, then it

Statpacks and Reporting Results

Principal components analysis is almost certainly a more straightforward proposition with your statpack than either of the other two approaches to multivariate analysis discussed here. Although a couple of other options are possible, principal components analysis almost always is based on correlation coefficients between variables. If you don't say otherwise, this is what the reader will assume. Even though little has been made here of the difference between principal components analysis and factor analysis, some readers will be much more concerned about the distinction, so it is important to pay attention to the options in your statpack and be sure to report accurately which you did. Providing the actual list of component loadings is an essential part of presenting the results. Just a description of the patterns observed on them is not enough.

As with multidimensional scaling, it is a good idea to limit the number of variables in a principal components analysis to no more than about half the number of cases. If the number of variables is much larger than this, there is substantial risk of finding spurious patterns that are no more than the product of random noise in the data.

really doesn't matter which you choose. If some details differ between the two, then perhaps these details should be taken with a grain of salt.

Principal components analysis is extremely powerful, just as regression analysis is. Much of this power comes from the specificity and rigidity of the model of relationships between variables, just as it does in regression analysis. Except for the possibility of inexplicable discrepancies between unrotated and rotated components, the results of principal components analysis tend to be clear and unequivocal. A principal components analysis will always produce components of some sort, whether there is much strong patterning in the dataset or not. In this sense it cannot fail, as multidimensional scaling can fail to produce any intelligible patterns. The signs that there just may not be much useful patterning in a dataset are easier to overlook in principal components analysis. If labeling the cases provides intuitive knowledge, then finding patterning in a multidimensional scaling is likely to be quite straightforward. On the other hand, if looking at the behavior of variables provides the most intuitive knowledge, as it does in the Ixcaquixtla household analysis, then presenting the multidimensional scaling in an intelligible way requires a good bit more work. With principal components analysis, it is just the reverse. If the behavior of variables provides the most direct route into identifying meaningful patterns, then principal components is certainly less work than multidimensional scaling because its natural way of presenting its results is organized around the variables.

Chapter 25
Cluster Analysis

Cluster analysis is perhaps the most familiar of all approaches to exploratory multivariate analysis, although it is not always thought of as a multivariate technique parallel to, for example, multidimensional scaling or principal components analysis. It is like such approaches, though, in seeking structure in the relationships among cases characterized by a number of variables. Cases that are strongly similar to each other, in terms of their values for a number of variables, wind up in the same groups or clusters, while those that are more different from each other wind up in different clusters. Cluster analysis mimics one of the human mind's fundamental ways of dealing with complicated variability: categorizing, or putting things into groups. Artifact typology in archaeology is a very familiar example of such categorizing. Recognizing that no two artifacts are likely to be identical, but that some pairs are more similar than others, we put the more similar ones together subjectively into what we then define as types. Our artifact typologies are *hierarchical* in that they group artifacts first according to broad classes like ceramics, flaked stone, textiles, etc., and then, within these broad classes, into more specific types at perhaps several levels. Flaked stone, for example, may be divided into tools and debitage; tools, in turn, into unifaces and bifaces; unifaces, into scrapers, blades, burins, etc.; scrapers, into endscrapers and sidescrapers; and so on.

This kind of hierarchical clustering can also be accomplished by statistical (as opposed to purely subjective) means. The first step in a hierarchical cluster analysis is usually the same as the first step in multidimensional scaling: measuring the similarities between each pair of cases in the dataset. The coefficients of similarity (or dissimilarity) that were discussed in Chapter 22 are just as suitable for clustering. Once the similarities (or dissimilarities or distances) have been measured, the clustering can begin.

R.D. Drennan, *Statistics for Archaeologists*, Interdisciplinary Contributions
to Archaeology, DOI 10.1007/978-1-4419-0413-3_25,
© Springer Science+Business Media, LLC 2004, 2009

Hierarchical cluster analysis is most often *agglomerative* because it usually proceeds by combining individual cases together to form larger and larger clusters. The procedure begins with each case considered as a separate entity. At the first step of this multistep procedure, the two most similar cases are combined into a cluster. At the next step, either two other cases are combined to begin a second cluster or else a third case is added to the existing cluster. Step by step, the clustering procedure continues to higher and higher (or more and more inclusive) levels in the hierarchy of agglomeration until finally all cases are combined into a single big cluster.

There are three major variations on this basic theme of hierarchical clustering, involving different specific *clustering criteria*. They arise because, as individual cases are gradually combined into more and more inclusive clusters, the procedure must choose between combining two individual cases to initiate a new cluster, adding a case to an existing cluster, or joining together two already existing clusters. While similarities have already been measured for all possible pairs of individual cases, the question arises of just how to use these in measuring the similarity between a case and an already existing cluster or between two already existing clusters. There may, for example, be a very strong similarity between a case in one cluster and a case in another cluster but very little similarity between the other pairs of cases involved in the two clusters.

SINGLE LINKAGE CLUSTERING

The simplest approach is single linkage clustering, in which the strongest single similarity score between cases governs each step in clustering. For example, using the matrix of similarity coefficients in Table 25.1, the sequence of single link clustering would go as follows:

1. The strongest single similarity in the matrix is 0.96 between Cases 4 and 6, so these two cases would be combined into a cluster.
2. The next strongest similarity in the matrix is 0.95 between Cases 1 and 3, so these two cases would be combined into a second cluster.

Table 25.1. Matrix of Similarity Coefficients for Seven Cases

	1	2	3	4	5	6	7
1	1.00						
2	0.34	1.00					
3	0.95	0.22	1.00				
4	0.69	0.04	0.11	1.00			
5	0.87	0.90	0.75	0.63	1.00		
6	0.12	0.15	0.37	0.96	0.27	1.00	
7	0.86	0.76	0.32	0.59	0.43	0.49	1.00

3. The next strongest similarity in the matrix is 0.90 between Cases 2 and 5, so these two cases would be combined into a third cluster.
4. The next strongest similarity in the matrix is 0.87 between Cases 1 and 5. Case 1 already belongs to a cluster (with Case 3) and Case 5 already belongs to a cluster (with Case 2), so these two clusters would be combined to form a cluster of four cases.
5. The next strongest similarity in the matrix is 0.86 between Cases 1 and 7. Case 1 already belongs to a cluster (with Cases 2, 3, and 5), so Case 7 would be added to that cluster, enlarging it to five cases.
6. The next strongest similarity in the matrix is 0.76 between Cases 7 and 2. These two already belong to the same cluster so there is no additional joining at this point.
7. The next strongest similarity in the matrix is 0.75 between Cases 3 and 5. These are also already members of the same cluster.
8. The next strongest similarity in the matrix is 0.69 between Cases 1 and 4. Case 1 already belongs to a cluster (with Cases 2, 3, 5, and 7), and Case 4 already belongs to a cluster (with Case 6), so these two clusters would be joined together to form a cluster of seven cases.

At this point all the cases have been joined into a single cluster, and the procedure is finished. The *dendrogram* in Fig. 25.1 provides a complete account of the entire procedure. The joining steps can be read from left to right. That is, the leftmost vertical line joining two cases represents the first joining step, and each step can be read in succession by selecting vertical joining lines in a steady rightward progression. The strength of similarity at any particular step in joining can be read from the horizontal scale at the top of the dendrogram.

Single linkage clustering often joins clusters together even though a number of individual cases from the two clusters show very little similarity to each other. That is, a single strong similarity for one pair of cases can cause two clusters to be joined even if all the other cases involved are quite different from each other. In the example used above, for instance, the cluster consisting of Cases 1 and 3 was joined with the cluster consisting of Cases 2 and 5 at Step 4 because of the strong similarity (0.85)

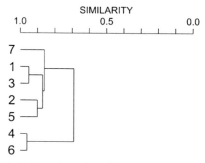

Figure 25.1. Single linkage clustering of the similarity scores from Table 25.1.

between Cases 1 and 5, even though the new cluster formed incorporated cases that were often not measured to be very similar. The similarity between Cases 1 and 2 is only 0.34 and between Cases 2 and 3 is only 0.22. Yet these pairs of cases were included together in the newly formed cluster. This would seem to make little sense, so clustering criteria other than single linkage are often used.

COMPLETE LINKAGE CLUSTERING

Complete linkage clustering prevents the undesirable joining of clusters with dissimilar members. With complete linkage, no two clusters are joined together unless even the weakest similarity between any two of the cases involved is stronger than any other "unused" similarity score in the matrix. The sequence of complete linkage clustering of the similarity scores from Table 25.1 would be as follows:

1.–3. The first three steps would be the same as with single linkage.
4. The next strongest similarity in the matrix is 0.87 between Cases 1 and 5. Case 1 already belongs to a cluster (with Case 3) and Case 5 already belongs to a cluster (with Case 2), so we would need to examine the other similarities involved in combining these two clusters. Cases 3 and 5 have a similarity of 0.75; Cases 1 and 2, of 0.34; and Cases 2 and 3, of 0.22. The two clusters will not be combined unless all the other possible combinations at this stage would require combining even less similar cases. All the possible combinations (and the weakest similarities each would incorporate) are as follows:

    1/3 with 7 (a similarity of 0.32 between 3 and 7);
    1/3 with 4/6 (a similarity of 0.11 between 3 and 4);
    1/3 with 2/5 (a similarity of 0.22 between 2 and 3);
    2/5 with 7 (a similarity of 0.43 between 5 and 7);
    2/5 with 4/6 (a similarity of 0.04 between 2 and 4); and
    4/6 with 7 (a similarity of 0.49 between 6 and 7).
    Combining 7 with the cluster of 4 and 6 puts together less dissimilar cases than any other available combination, so step 4 makes the cluster 4/6/7.

5. At this point there are three clusters, made up of Cases 1/3, 2/5, and 4/6/7. Cluster 1/3 is held back from joining Cluster 2/5 by a weak similarity of 0.22 between Cases 2 and 3. The only other two possible joining steps, however, are impeded by even weaker similarities. The similarity of 0.11 between Cases 3 and 4 holds Clusters 1/3 and 4/6/7 apart, and the similarity of 0.04 between Cases 2 and 4 holds Clusters 2/5 and 4/6/7 apart. Thus, the procedure would work its way down to the similarity of 0.22 as the next strongest combination, and would put Clusters 1/3 and 2/5 together as the fifth step.
6. The last joining would unite Cluster 1/2/3/5 with Cluster 4/6/7 at the weakest level of 0.04 (the similarity between Cases 2 and 4).

This history of clustering and its outcome are represented in the dendrogram of Fig. 25.2. Complete linkage clustering can be seen to represent the opposite extreme

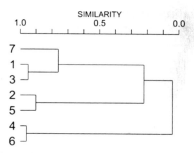

Figure 25.2. Complete linkage clustering of the similarity scores from Table 25.1.

to single linkage clustering. Instead of relying on the single strongest link between clusters (as with single linkage), it relies on the single weakest link to determine what to join. The effects of the two criteria can be appreciated by imagining each cluster as a club whose members are deciding which applicants to accept. The effect of complete linkage clustering is the same as the system of "blackballing" applicants for membership. Any case can prevent the joining of any other case with which it has a very low similarity, just as in a club where any single member can veto a disliked applicant. Single linkage clustering, however, is just the opposite: a new member is accepted into the club on the basis of a strong similarity score from any current member, regardless of how low other members' scores for that applicant may be.

AVERAGE LINKAGE CLUSTERING

Just as clubs may adopt procedures for accepting new members that fall somewhere between the two extremes, average linkage clustering has been proposed as a happy medium between single and complete linkage. In average linkage clustering, after each joining step a new matrix of similarities is calculated, treating each existing cluster as if it were a single case. The similarity between an existing cluster and another case is the average of the similarities between that case and each member of the cluster. For the similarity coefficients from Table 25.1, average linkage clustering would proceed as follows:

1. Cases 4 and 6, with a similarity of 0.96, would join. The similarity matrix would be recalculated, with the result in Table 25.2. The similarity score between Cluster 4/6 and Case 1, for example, would be the average of the similarity scores between Cases 1 and 4 and Cases 1 and 6, or: $0.69 + 0.12/2 = 0.41$.
2. Cases 1 and 3, with a similarity of 0.95, would combine, and the similarity matrix would be recalculated again, with the result in Table 25.3. The similarity between Clusters 1/3 and 4/6 would be the average of the similarities between the four pairs of cases involved (1 and 4, 1 and 6, 3 and 4, and 3 and 6).
3. Cases 2 and 5, with a similarity of 0.90, would combine, and the similarity matrix would be recalculated again, with the result in Table 25.4. (The first three steps,

Table 25.2. Matrix of Similarity Coefficients after the First Step in Average Link Clustering

	1	2	3	4/6	5	7
1	1.00					
2	0.34	1.00				
3	0.95	0.22	1.00			
4/6	0.41	0.10	0.24	1.00		
5	0.87	0.90	0.75	0.45	1.00	
7	0.86	0.76	0.32	0.54	0.43	1.00

Table 25.3. Matrix of Similarity Coefficients after the Second Step in Average Link Clustering

	1/3	2	4/6	5	7
1/3	1.00				
2	0.28	1.00			
4/6	0.32	0.10	1.00		
5	0.81	0.90	0.45	1.00	
7	0.59	0.76	0.54	0.43	1.00

Table 25.4. Matrix of Similarity Coefficients after the Third Step in Average Link Clustering

	1/3	2/5	4/6	7
1/3	1.00			
2/5	0.56	1.00		
4/6	0.32	0.27	1.00	
7	0.59	0.60	0.54	1.00

Table 25.5. Matrix of Similarity Coefficients after the Fourth Step in Average Link Clustering

	1/3	2/5/7	4/6
1/3	1.00		
2/5/7	0.56	1.00	
4/6	0.32	0.36	1.00

Table 25.6. Matrix of Similarity Coefficients after the Fifth Step in Average Link Clustering

	1/2/3/5/7	4/6
1/2/3/5/7	1.00	
4/6	0.35	1.00

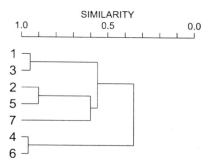

Figure 25.3. Average linkage clustering of the similarity scores from Table 25.1.

then, turn out to be the same for average linkage as they were for both single and complete linkage.)

4. Case 7 would join with Cluster 2/5, based on their similarity score of 0.60, and the similarity matrix would be recalculated yet again, with the result in Table 25.5.
5. The highest similarity score in the new matrix, 0.56, would cause Cluster 1/3 to join with Cluster 2/5/7, and the similarity matrix would be recalculated one last time, with the result in Table 25.6.
6. The final two clusters would be joined together at a similarity level of 0.35.

This clustering sequence and its outcome are illustrated in the dendrogram in Fig. 25.3. Minor variations in average linkage clustering can be produced by calculating the new similarity matrix in different ways – by using the median similarity coefficient instead of the average, for example, or by calculating some other index of the center of the scores involved (sometimes referred to as a centroid).

WHICH LINKAGE CRITERION TO CHOOSE

Comparison of the three dendrograms in Figs. 25.1–25.3 shows the different consequences of the three different linkage criteria. Sometimes the difference is quite dramatic. This naturally raises the question of which linkage criterion should be chosen. There is no simple principle that can be routinely applied to answer this question.

Occasionally the nature of the dataset suggests a particular linkage criterion as the most appropriate. For example, hierarchical clustering is frequently used in raw material sourcing studies. Here the cases are typically artifacts that have been subjected to some form of chemical analysis, and the variables are some measure of the abundance of different chemical elements or other constituents. Hierarchical clustering may be used to delineate groups of artifacts that are presumed to be made of materials from the same source location. If we expect the raw material involved to have pretty much the same composition in a given source, then complete linkage clustering makes good sense. This is because complete linkage clustering will prevent assigning an artifact to a cluster unless it is broadly similar to each of the artifacts already in the cluster. If we hope for clusters whose basis is shared raw material source, then we should insist, just as complete linkage clustering does, that each member of the group be similar to each other member. In fact, complete linkage clustering often does produce very convincing results in raw material sourcing studies.

The most important concern really is whether a particular linkage criterion produces interpretable results. The aim is to produce clusters that make good sense according to whatever considerations can be applied to their interpretation. If one linkage criterion produces sensible clusters, then that is the best solution, even if there were good a priori reasons to think that a different linkage criterion might work well with that dataset. Fortunately, once similarity scores are calculated satisfactorily, it is extremely easy to get a statpack to produce hierarchical clusterings based on different criteria and compare the results.

HOW MANY CLUSTERS TO DEFINE

Since hierarchical clustering starts with each case in a separate cluster and finishes with all cases in a single large cluster, it is necessary to decide how many clusters to read out of the results. This is, in effect, a process of deciding where to stop the clustering procedure for purposes of interpretation. If we consider the average linkage clustering in the example above to have run its useful course after Step 5, then two clusters are produced (1/2/3/5/7 and 4/6). If, on the other hand, we find meaning in the three clusters 1/3, 2/5/7, and 4/6, we could cut the process off after Step 4. This is not an option that must be set when running the analysis, but instead a question of reading the results. As with the choice of the number of dimensions in a multidimensional scaling or the number of components in a principal components analysis, though, the decision devolves primarily on what is meaningful. For this, there are really no rules. It depends on the prior knowledge, intuition, and inventiveness of the analyst.

CLUSTERING BY VARIABLES

Usually hierarchical clustering is done by cases, and that is what all of the above concerns. Occasionally, however, a hierarchical clustering of variables can be useful and enlightening. The starting point for a hierarchical clustering of variables is to measure the similarities between variables. Thinking about the similarities between variables requires a shift of mental gears after thinking about similarities between cases. Similarities between variables amount to a consideration of how similarly the variables behave across cases. Two variables that vary together across the cases in the dataset are quite similar. If the variables are all measurements, correlation

---

Statpacks and Reporting Results

Cluster analysis, like multidimensional scaling, is performed by many but not all large statpacks. The measurement of similarities, whether between variables or between cases, sometimes appears as an option to be selected as part of the clustering routines. It is conceptually a separate step, however, and more options for different measures of similarity may be available elsewhere in your statpack or in stand-alone programs. Either way, it is important in reporting results to be explicit about the exact nature of the variables, the similarity coefficient used, and the clustering criterion selected. The resulting dendrogram is the essential result to present to back up your discussion of what you've found out.

As with multidimensional scaling and principal components analysis, it is a good idea to limit the number of variables in a cluster analysis to no more than about half the number of cases. If the number of variables is much larger than this, there is substantial risk of finding spurious patterns that are no more than the product of random noise in the data.

---

coefficients (r) are a good measure of similarity between variables. This is, after all, the starting point for principal components analysis, which is also an analysis by variables. In many instances it may make most sense to use the absolute values of correlations, rather than the values with signs. This is because a strong negative correlation may be just as meaningful a similarity between two variables as a strong positive correlation. The dissimilar variables, then, would be those that show little relationship to each other at all. The decision about whether to use correlations or absolute values of correlations depends on the specific context and content of the variables in a particular dataset.

If the variables are not all measurements, but include ranks or presence/absence variables, correlation coefficients may still be a perfectly reasonable choice, for the same reasons that such variables often turn out to create no real obstacles to principal components analysis. If there are variables with multiple unranked categories, then the use of correlation coefficients is not appropriate. If all the variables are categorical, we can use V or ϕ or ϕ^2, the measures of strength that accompany the chi-square test (Chapter 14).

Either way, the starting point for the clustering is a square symmetric matrix of scores (r or V values) of each variable with each other variable. The same measure must be chosen for the entire matrix, however, since r and V values are not really comparable. This matrix is handled in the clustering procedure exactly as it would be if it measured similarities of each case with each other case. A linkage criterion must be chosen, and clustering can proceed.

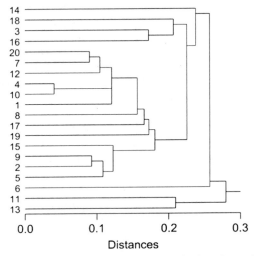

Figure 25.4. Clustering of the household units from Ixcaquixtla.

CLUSTERING THE IXCAQUIXTLA HOUSEHOLD DATA

The starting point for a clustering by cases of the household units from Ixcaquixtla is the matrix of similarity coefficients in Table 22.9. The dendrogram in Fig. 25.4 shows a single-linkage clustering. It might be read in several ways. Sometimes a small grouping can be found that makes some sense to work from. Household Units 15, 9, 2, and 5, for example, form a cluster sharing the large amounts of energy invested in burials and the high proportions of bowls, decorated ceramics, and faunal remains previously interpreted in this dataset as associated with wealth. In the other clearest cluster, Household Units 20, 7, and 1 have high proportions of debitage, although Household Units 12, 4, and 10 do not share this characteristic even though they also appear in this cluster. All six, however, seem to be of very similar modest wealth. Units 6, 11, and 13 are the households with high proportions of obsidian. Both households 8 and 17 have high proportions of kiln wasters. Households 11 and 13 have high proportions of both kiln wasters and obsidian. Household Units 14 and 18 have mace heads in burials and the houses are on platforms, but they do not join well, even though they turn up next to each other in the list. Households 3 and 11, which also have both these things, are widely separated. Some of the patterns seen before in this dataset are, thus, visible here, although mostly in fragments rather than in integrated ways.

This is the least satisfactory of all the analyses we have tried so far on this dataset. The reason is clear. The nature of the patterns we have seen in the Ixcaquixtla household data is just not at all well expressed by simply grouping the household units together into several mutually exclusive clusters, but this is the only kind of pattern hierarchical clustering can delineate. We probably would expect a dataset like this to

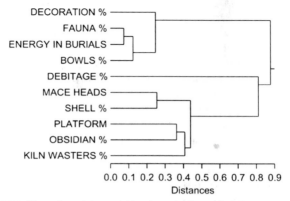

Figure 25.5. Clustering of the variables from the Ixcaquixtla household dataset.

have more complicated patterns that, if thought of as groups of households, would involve patterns of multiple overlapping and crosscutting groups. If we expected that we would need to recognize patterns of this sort, we would not choose hierarchical clustering as an analytical approach. Hierarchical clustering does provide an effective solution, however, when mutually exclusive groupings accurately characterize the patterns we need to find. This is why it has proven to be especially useful in questions of raw material sourcing. In raw material sourcing, each artifact is expected to fit comfortably into one and only one group which represents a distinct raw material source, and this is precisely the kind of pattern hierarchical clustering inevitably produces.

A clustering by variables of the Ixcaquixtla household data (Fig. 25.5) is only slightly more satisfactory than the clustering by cases. The similarities between variables for this analysis were measured with correlation coefficients (r), so the starting point was the same as for the principal components analysis. There is a clear cluster of the four variables that formed a gradient in Dimensions 1 and 2 of the multidimensional scaling and that loaded very strongly on the first component in the principal components analysis. Proportion of lithic debitage does not join with any other variable until very near the end of the clustering, and this is consistent with what we have been seeing through several analyses. The pattern of clustering between the other five variables though is less satisfactory. We have seen relationships between the presence of mace heads and high proportions of marine shell, so it makes sense that these two variables cluster together. Platform, however, does not appear in this cluster but rather in a cluster with proportion of obsidian. That is a relationship we have seen before, but then obsidian's connection to kiln wasters pulls that cluster together with kiln wasters rather than with the shell and mace head cluster. Again the complexity of crosscutting relationships has confounded the dendrogram structure created by hierarchical clustering. Just as with cases, when variables really need to belong to multiple groups, clustering is likely to be predicated on too simple a structure of relationships to give very satisfactory results.

The dataset we have been using as an example, then, is one whose structure just does not match the only kind of structure that hierarchical clustering can find. In this sense, hierarchical cluster analysis is the narrowest of the approaches we have considered. Both multidimensional scaling and principal components analysis offer a wider view of patterning in a multivariate dataset and are thus applicable in a broader range of contexts. When a clear and simple structure of mutually exclusive groupings is indicated, though, hierarchical clustering is probably the approach of choice. It will delineate such clusters more clearly and effectively than either multidimensional scaling or principal components analysis.

In the end, the approaches to multivariate analysis described in these final chapters are tools for exploring multivariate datasets, looking for patterning both suspected and unsuspected. It often pays to try out different approaches and different options within each approach (including different similarity coefficients for multidimensional scaling and clustering). It is often not possible to predict in advance which approach will work well with a particular dataset, and experimenting with different ones can produce insights that would not be obtained in any other way.

Suggested Reading

Bibliographic citations have been avoided in this book in order to streamline the presentation and because careful tracing of the intellectual pedigree of many of the ideas and techniques discussed here is a scholarly endeavor in itself, and one not comfortably combined with an introduction to their application in archaeology. The books and articles listed below, however, are places to go for further information on statistics in archaeology. The literature on statistics in archaeology has become very large, and the list below is very short. Consequently, a large number of perfectly relevant references have not been included – the selection is idiosyncratic rather than comprehensive. Some of the items included are relatively new; some are not so new. Some are included because they share the general outlook of this book (and indeed in some cases are the specific inspiration for it); some, because they complement it (which is to say they take a different perspective).

GENERAL STATISTICS BOOKS

Exploratory Data Analysis, by John W. Tukey (Reading, MA: Addison-Wesley, 1977), is one of the classic presentations of an approach to statistics from which much in this book is derived, and its author is the father of the approach. Not surprisingly, there is a great deal more to exploratory data analysis (EDA) than has been presented in this volume, and readers who would like to go directly to the source to find out about it should read Tukey's book, which is a full-scale introductory text in EDA. Although EDA is now more than 40 years old, only parts of the prescription Tukey laid out for EDA have been much applied in archaeology (and even those parts that have been do not yet constitute the "standard" statistical approach in the archaeological literature). Many of the techniques Tukey discusses in his book were intended to be easily accomplished with pencil and paper, or, at most, with a calculator, but more widespread availability in the most commonly used computer statpacks would undoubtedly encourage greater use of EDA techniques in archaeology and in other fields.

Exploratory Data Analysis, by Frederick Hartwig and Brian E. Dearing (Beverly Hills, CA: Sage, 1979), is a brief presentation of the basic techniques of EDA. It nevertheless includes a number of EDA topics not covered in this volume.

Applications, Basics, and Computing of Exploratory Data Analysis, by Paul F. Velleman and David C. Hoaglin (Boston, MA: Duxbury Press, 1981), is yet another introduction to EDA techniques, less formidable than Tukey's and more comprehensive than Hartwig and Dearing's.

Understanding Data, by Bonnie H. Erikson and T. A. Nosanchuk (Toronto: McGraw-Hill Ryerson, 1977), is an introductory statistics text that combines EDA with more traditional statistical approaches. It advocates two different and complementary kinds of work with numbers (exploratory and confirmatory), keeping the two strongly separated and emphasizing the differences between their goals. The presentation is especially accessible and free of jargon and abstract mathematics.

Introduction to Contemporary Statistical Methods, by Lambert H. Koopmans (Boston, MA: Duxbury Press, 1987), also combines EDA with more traditional statistics. A very wide range of methods is covered, and the logic behind the methods is presented in more abstract mathematical terms than in most of the other books listed here. Instead of focusing on the difference between exploration and confirmation throughout the book, Koopmans considers statistical exploration at the beginning, and then complements the discussion of the usual significance testing techniques with a wide array of robust techniques suitable for use on data that present problems for the usual techniques.

Nonparametric Statistics for the Behavioral Sciences, by Sidney Siegel (New York: McGraw-Hill, 1956), is a classic presentation of a full array of robust techniques for evaluating significance – that is, ones that are not much affected by things like very asymmetrically shaped batches for which means and standard deviations are not useful. Many of these techniques require special tables in which to look up the results, and Siegel provides them.

Sampling Techniques, by William G. Cochran (New York: Wiley, 1977), describes itself (quite accurately) as "a comprehensive account of sampling theory." It is, perhaps, the ultimate source on this subject. Estimating means and proportions, sample selection, stratified sampling, cluster sampling, sampling with and without replacement, determining necessary sample size, and many other topics are covered in detail. The full logic behind the techniques presented is given in mathematical terms.

Elementary Survey Sampling, by Richard L. Scheaffer, William Mendenhall, and Lyman Ott (Boston, MA: Duxbury Press, 1986), covers much of the same ground that Cochran does. The presentation is largely in terms of abstract mathematics, but it is considerably less detailed and formidable than Cochran's.

INTRODUCTIONS TO STATISTICS FOR (AND OFTEN BY) ARCHAEOLOGISTS

Sampling in Archaeology, by Clive Orton (Cambridge: Cambridge University Press, 2000), explores sampling in archaeology at length. The emphasis is on sensible use of sampling theory in the varied array of circumstances archaeological data collection presents. There is especially extended treatment of sampling in the field, which usually means spatially based sampling. Examples are drawn from real archeological datasets.

Quantifying Archaeology, by Stephen Shennan (Edinburgh: Edinburgh University Press, 1997), is an introductory statistics text (and more) specifically for archaeologists. Mostly traditional statistical methods are covered, but some EDA techniques are also included. Shennan goes beyond basic statistical principles to deal with multivariate analysis (with emphasis on multiple regression, clustering, and principal components and factor analysis). Methods for estimating population means and proportions are presented (not usual in introductory statistics books) and the special issues that sampling raises in archaeology are discussed.

Statistics in Archaeology, by Michael Baxter (London: Arnold, 2003), reviews basic statistical techniques (including the fundamental ones from EDA) very compactly, and considers a number of special topics from an archaeological perspective. These include several multivariate approaches, spatial analysis, radiocarbon dating, seriation, and assemblage diversity. Baxter's earlier book, *Exploratory Multivariate Analysis in Archaeology* (Edinburgh: Edinburgh University Press, 1994), considers multivariate analysis in greater depth. Extended treatment is given to principal components analysis, correspondence analysis, cluster analysis, and discriminant analysis, and numerous examples of multivariate analyses of real archaeological data are woven into the explanations.

Digging Numbers: Elementary Statistics for Archaeologists, by Mike Fletcher and Gary R. Lock (Oxford: Oxford University Committee for Archaeology, 2005), applies basic statistical techniques (both traditional and EDA) specifically to archaeology. The presentation is informal, avoids jargon, and is designed to be very accessible, especially to those suffering math anxiety.

Refiguring Anthropology: First Principles of Probability and Statistics, by David Hurst Thomas (Prospect Heights, IL: Waveland Press, 1986), is an introductory statistics text specifically for anthropologists (including archaeologists). The approach is purely traditional (that is, it does not incorporate an EDA perspective or techniques), and some rules are laid down that this volume has argued against, but numerous robust methods are discussed. There are abundant examples of the application of all the techniques presented to real data from archaeology, cultural anthropology, and biological anthropology.

ARCHAEOLOGISTS CONSIDER STATISTICS IN OUR DISCIPLINE

"The Trouble with Significance Tests and What We Can Do About It," by George L. Cowgill (*American Antiquity* 42:350–368, 1977), makes the case for an attitude about significance testing that has inspired much in the perspective taken on this subject in this volume. It is a view distinctly different from that often adopted in introductory statistics texts – indeed it is branded as heresy by the rules often found in introductory statistics texts. This article is fundamental for those interested in a fuller presentation of the arguments that archaeologists will often find it useful to use samples directly to make estimates about populations and that it is usually a mistake for archaeologists to force significance tests into the mold of a yes-or-no decision. Cowgill's suggestions about the most useful ways to approach these issues in archaeology go well beyond what is presented in this volume, which has stopped at the point where the information commonly provided by computer statpacks imposes a limitation.

"A Selection of Samplers: Comments on Archaeo-statistics," by George L. Cowgill [In *Sampling in Archaeology*, edited by James W. Mueller (Tucson: University of Arizona Press, 1975)], prefigures some of the issues Cowgill argues more fully in his later paper (above), and focuses especially on sampling, criticizing many of what he sees as erroneous notions that appear in other papers in the same volume.

"On the Structure of Archaeological Data," by Mark S. Aldenderfer [In *Quantitative Research in Archaeology: Progress and Prospects*, edited by Mark S. Aldenderfer (Newbury Park, CA: Sage, 1987)], is a discussion of the fundamental nature of data in archaeology, the position that numbers occupy in such data, and the implications that this has for how we think about and analyze data. Four other articles in this same volume are also of special interest. "Quantitative Methods Designed for Archaeological Problems," by Keith W. Kintigh, discusses the issue of the extent to which standard statistical techniques and those borrowed directly from other disciplines are suited to the particular needs of archaeology. "Simple Statistics," by Robert Whallon, stresses the importance of exploring the patterns in numbers in batches before proceeding to more complicated analyses. And "Archaeological Theory and Statistical Methods: Discordance, Resolution, and New Directions," by Dwight W. Read, and "Removing Discordance from Quantitative Analysis," by Christopher Carr, try to place archaeological data analysis firmly in a broader context. Both authors are concerned that data analysis is too often conceived and carried out in isolation from the theoretical questions that analysis aims to help answer. As a consequence "discordance" between data, analysis, and theory arises and seriously impedes the archaeological endeavor.

"Statistics for Archaeology," also by Aldenderfer [In *Handbook of Archaeological Methods*, edited by Herbert G.D. Machsner and Christopher Chippindale (Lanham, MD: Altamira Press, 2005)], reviews the history of statistical analysis in archaeology and considers a series of topics of special importance in archaeology, including spatial analysis, EDA techniques, and Bayesian analysis.

MULTIVARIATE ANALYSIS

Since the chapters here on multivariate analysis are only brief explanations, further reading on these techniques is likely to be especially important. Several of the volumes listed above on statistics in archaeology also provide introductory treatments, and some are more detailed than those in the final chapters of this book. In addition, the sources below may be helpful.

In *Multidimensional Scaling* (Quantitative Applications in the Social Sciences, Paper 11, Beverly Hills, CA: Sage, 1972), Joseph B. Kruskal and Myron Wish provide a clear introduction to this multivariate approach. Kruskal also wrote an article directed specifically at archaeologists ["Multi-Dimensional Scaling in Archaeology: Time Is Not the Only Dimension." In *Mathematics in the Archaeological and Historical Sciences*, edited by F.R. Hodson, D.G. Kendall, and P. Tautu, eds., Edinburgh: Edinburgh University Press, 1971]. *Multidimensional Scaling: Theory and Applications in the Behavioral Sciences*, edited by Roger N. Shepard, A. Kimball Romney, and Sara Beth Nerlove (2 volumes, New York: Seminar Press, 1972), is a compilation of studies making use of multidimensional scaling in all parts of the social sciences. Ingwer Borg and Patrick J.F. Groenen bring the subject more up to date in *Modern Multidimensional Scaling: Theory and Applications* (New York: Springer, 1997).

Principal Components Analysis, by George H. Dunteman (Quantitative Applications in the Social Sciences, Paper 69. Newbury Park, CA: Sage, 1989), is an accessible but comprehensive account, with numerous examples. Two other volumes in the same series of publications, both by Jae-On Kim and Charles W. Mueller, discuss factor analysis, as opposed to principal components analysis, in fairly intuitive terms. They are *Introduction to Factor Analysis: What It Is and How to Do It* (Quantitative Applications in the Social Sciences, Paper 13. Beverly Hills, CA: Sage, 1978) and *Factor Analysis: Statistical Methods and Practical Issues* (Quantitative Applications in the Social Sciences, Paper 14. Beverly Hills, CA: Sage, 1978).

A very accessible discussion of measures of similarity and basic principles of cluster analysis can be found in *Cluster Analysis* by Mark S. Aldenderfer and Roger K. Blashfield (Quantitative Applications in the Social Sciences, Paper 44. Newbury Park, CA: Sage, 1984). *Numerical Taxonomy*, by Peter H.A. Sneath and Robert R. Sokal (San Francisco, CA: Freeman, 1973), is one of the classic sources on hierarchical clustering. *Cluster Analysis for Applications*, by Michael R. Anderberg (New York: Academic, 1973), is another. Both these books also address the measurement of similarities between cases at length.

Index

Lightning Source UK Ltd.
Milton Keynes UK
10 September 2010

159734UK00001B/20/P